This report contains the collective views of an international group of experts and does not necessarily represent the decisions or the stated policy of the United Nations Environment Programme, the International Labour Organisation, or the World Health Organization

Environmental Health Criteria 85

LEAD –

ENVIRONMENTAL ASPECTS

Published under the joint sponsorship of
the United Nations Environment Programme,
the International Labour Organisation,
and the World Health Organization

World Health Organization
Geneva, 1989

The **International Programme on Chemical Safety (IPCS)** is a joint venture of the United Nations Environment Programme, the International Labour Organisation, and the World Health Organization. The main objective of the IPCS is to carry out and disseminate evaluations of the effects of chemicals on human health and the quality of the environment. Supporting activities include the development of epidemiological, experimental laboratory, and risk-assessment methods that could produce internationally comparable results, and the development of manpower in the field of toxicology. Other activities carried out by the IPCS include the development of know-how for coping with chemical accidents, coordination of laboratory testing and epidemiological studies, and promotion of research on the mechanisms of the biological action of chemicals.

ISBN 92 4 154285 3

The designations employed and the presentation of the material in this publication do not imply the expression of any opinion whatsoever on the part of the Secretariat of the World Health Organization concerning the legal status of any country, territory, city or area or of its authorities, or concerning the delimitation of its frontiers or boundaries.

The mention of specific companies or of certain manufacturers' products does not imply that they are endorsed or recommended by the World Health Organization in preference to others of a similar nature that are not mentioned. Errors and omissions excepted, the names of proprietary products are distinguished by initial capital letters.

ISSN 0250-863X

PRINTED IN FINLAND

89/7935 — VAMMALA — 6500

CONTENTS

WHO TASK GROUP ON ENVIRONMENTAL HEALTH CRITERIA FOR LEAD - ENVIRONMENTAL ASPECTS

Members

Dr L.A. Albert, Environmental Pollution Programme, National Institute for Research on Biotic Resources, Veracruz, Mexico
Dr R. Elias, Environmental Criteria and Assessment Office, US Environmental Protection Agency, Research Triangle Park, North Carolina, USA *(Chairman)*
Dr J.H.M. Temmink, Department of Toxicology, Agricultural University, Biotechnion, Wageningen, Netherlands
Dr G. Roderer, Fraunhofer Institute for Environmental Chemistry and Ecotoxicology, Schmallenberg-Grafschaft, Federal Republic of Germany
Dr R. Koch, Division of Toxicology, Research Institute for Hygiene and Microbiology, Bad Elster, German Democratic Republic
Dr Y. Kodama, Department of Environmental Health, University of Occupational and Environmental Health, Kitakyushu, Japan
Professor P.N. Viswanathan, Ecotoxicology Section, Industrial Toxicology Research Centre, Lucknow, India

Observers

Mr D.J.A. Davies, Department of the Environment, London, United Kingdom
Dr I. Newton, Institute of Terrestrial Ecology, Monks Wood Experimental Station, Huntingdon, United Kingdom

Secretariat

Dr S. Dobson, Institute of Terrestrial Ecology, Monks Wood Experimental Station, Huntingdon, United Kingdom *(Rapporteur)*
Dr M. Gilbert, International Programme on Chemical Safety, World Health Organization, Geneva, Switzerland *(Secretary)*
Mr P.D. Howe, Institute of Terrestrial Ecology, Monks Wood Experimental Station, Huntingdon, United Kingdom

NOTE TO READERS OF THE CRITERIA DOCUMENTS

Every effort has been made to present information in the criteria documents as accurately as possible without unduly delaying their publication. In the interest of all users of the environmental health criteria documents, readers are kindly requested to communicate any errors that may have occurred to the Manager of the International Programme on Chemical Safety, World Health Organization, Geneva, Switzerland, in order that they may be included in corrigenda, which will appear in subsequent volumes.

*   *   *

A detailed data profile and a legal file can be obtained from the International Register of Potentially Toxic Chemicals, Palais des Nations, 1211 Geneva 10, Switzerland (Telephone No. 988400 - 985850).

ENVIRONMENTAL HEALTH CRITERIA FOR LEAD - ENVIRONMENTAL ASPECTS

A WHO Task Group on Environmental Health Criteria for Lead - Environmental Aspects met at the Institute of Terrestrial Ecology, Monks Wood, United Kingdom, from 7 to 11 December 1987. Dr B.N.K. Davis welcomed the participants on behalf of the host institution, and Dr M. Gilbert opened the meeting on behalf of the three co-sponsoring organizations of the IPCS (ILO/UNEP/WHO). The Task Group reviewed and revised the draft criteria document and made an evaluation of the risks for the environment from exposure to lead.

The first draft of this document was prepared by Dr S. Dobson and Mr P.D. Howe, Institute of Terrestrial Ecology. Dr M. Gilbert and Dr P.G. Jenkins, both members of the IPCS Central Unit, were responsible for the overall scientific content and editing, respectively.

* * *

Partial financial support for the publication of this criteria document was kindly provided by the United States Department of Health and Human Services, through a contract from the National Institute of Environmental Health Sciences, Research Triangle Park, North Carolina, USA - a WHO Collaborating Centre for Environmental Health Effects.

INTRODUCTION

There is a fundamental difference in approach between the toxicologist and the ecotoxicologist concerning the appraisal of the potential threat posed by chemicals. The toxicologist, because his concern is with human health and welfare, is preoccupied with any adverse effects on individuals, whether or not they have ultimate effects on performance or survival. The ecotoxicologist, in contrast, is concerned primarily with the maintenance of population levels of organisms in the environment. In toxicity tests, he is interested in effects on the performance of individuals - in their reproduction and survival - only insofar as these might ultimately affect the population size. To him, minor biochemical and physiological effects of toxicants are irrelevant if they do not, in turn, affect reproduction, growth, or survival.

It is the aim of this document to take the ecotoxicologist's point of view and consider effects on populations of organisms in the environment. No attempt has been made to link the conclusions reached in this document with possible effects on human health, since a new Environmental Health Criteria document examining the effects on human health of lead compounds is in preparation. Due attention has been given to persistence in the environment and bioaccumulation. These will have implications for human consumption of the metal.

This document, although based on a thorough survey of the literature, is not intended to be exhaustive in the material included. In order to keep the document concise, only those data which were considered to be essential in the evaluation of the risk posed by lead to the environment have been included. Concentration figures for lead in the environment, or in particular species of organism, have not been included unless they illustrate specific toxicological points. "Snap shot" concentration data, where a causal relationship between the presence of the metal and an observed effect is not clearly demonstrated, have been excluded.

The term bioaccumulation indicates that organisms take up chemicals to a greater concentration than that found in their environment or their food. "Bioconcentration factor" is a quantitative way of expressing bioaccumulation: the ratio of the concentration of the chemical in the organism to the concentration of the chemical in the environment or food. Biomagnification refers, in this document, to the progressive accumulation of chemicals along a food chain.

1. SUMMARY AND CONCLUSIONS

1.1 Physical and Chemical Properties and Sources of Pollution

Lead is a bluish or silvery-grey soft metal. With the exception of the nitrate, the chlorate, and, to a much lesser degree, the chloride, the salts of lead are poorly soluble in water. Lead also forms stable organic compounds. Tetraethyllead and tetramethyllead are used extensively as fuel additives. Both are volatile and poorly soluble in water. Trialkyllead compounds are formed in the environment by the breakdown of tetraalkylleads. These trialkyl compounds are less volatile and more readily soluble in water. Lead is mined, most usually as the sulfide, "galena". Pollution of the environment occurs through the smelting and refining of lead, the burning of petroleum fuels containing lead additives and, to a lesser extent, the smelting of other metals and the burning of coal and oil. Metallic lead deriving from shotgun cartridges or used as fishing weights is lost in the environment and often remains available to organisms.

1.2 Uptake, Loss, and Accumulation in Organisms

Lead in the environment is strongly adsorbed onto sediment and soil particles reducing its availability to organisms. Because of the low solubility of most of its salts, lead tends to precipitate out of complex solutions.

1.2.1 Model ecosystems

In aquatic and aquatic/terrestrial model ecosystems, uptake by primary producers and consumers seems to be determined by the bioavailability of the lead. Bioavailability is generally much lower whenever organic material, sediment, or mineral particles (e.g., clay) are present. In many organisms, it is unclear whether lead is adsorbed onto the organism or actually taken up. Consumers take up lead from their contaminated food, often to high concentrations, but without biomagnification.

1.2.2 Uptake and accumulation by aquatic organisms

The uptake and accumulation of lead by aquatic organisms from water and sediment are influenced by various environmental factors such as temperature, salinity, and pH, as well as humic and alginic acid content.

In contaminated aquatic systems, almost all of the lead is tightly bound to sediment. Only a minor fraction is dissolved in the water, even interstitial water between the sediment particles.

The lead uptake by fish reaches equilibrium only after a number of weeks of exposure. Lead is accumulated mostly in gill, liver, kidney, and bone.

Fish eggs show increasing lead levels with increased exposure concentration, and there are indications that lead is present on the egg surface but not accumulated in the embryo.

In contrast to inorganic lead compounds, tetraalkyllead is rapidly taken up by fish and rapidly eliminated after the end of the exposure.

1.2.3 Uptake and accumulation by terrestrial organisms

In bacteria, the majority of lead is associated with the cell wall. A similar phenomenon is also noted in higher plants. Some lead that passes into the plant root cell can be combined with new cell wall material and subsequently removed from the cytoplasm to the cell wall. Of the lead remaining in the root cell, there is evidence of very little translocation to other parts of the plant because the concentration of lead in shoot and leaf tissue is usually much lower than in root. Foliar uptake of lead occurs, but only to a very limited extent.

In animals, there is a positive correlation between tissue and dietary lead concentrations, although tissue concentrations are almost always lower. The distribution of lead within animals is closely associated with calcium metabolism.

Lead shot is typically trapped in the gizzard of birds where it is slowly ground down resulting in the release of lead.

The tetravalent organic form of lead is generally more toxic than the divalent, inorganic form, and its distribution in organisms may not specifically follow calcium metabolism.

1.2.4 Uptake of lead in the field

Organisms have been found to incorporate lead from the environment, generally in proportion to the degree of contamination. Lead deposition in a region depends on the air concentrations of the metal, which decrease with the distance from the source.

In shellfish, lead concentrations are higher in the calcium-rich shell than in the soft tissue; they relate to the concentrations in sediment.

Lead concentrations in some marine fish are higher in gills and skin than in other tissues, but this may be largely due to adsorption. Liver levels increase significantly with age.

In dolphins, lead is transferred from mothers to offspring during fetal development and lactation. This might be related to the calcium metabolism.

1.2.5 Uptake in the vicinity of highways and in urban areas

Lead concentrations are highest in soils and organisms close to roads where traffic density is high. The lead measured is inorganic and derives almost exclusively from alkyllead compounds added to petrol.

The lead in the soil and in vegetation decreases exponentially with the distance from the road. Lead is also found in the sediments of streams in the vicinity of highways.

Lead contamination increases lead levels in plants and animals in areas close to roads. These levels are positively correlated with traffic volume and proximity of roads.

Most lead deposited is found within 500 m of the road and within the upper few centimetres of soil. It can be assumed that lead levels in soil and biota are not influenced by traffic at distances from roads greater than this.

1.2.6 Uptake of lead from industrial sources

Terrestrial and aquatic plants accumulate lead in industrially contaminated environments. In aquatic plant species, lead uptake can occur from both water and sediment, although uptake from sediment usually predominates. Lead levels decrease with distance from the source and are lowest during the active growing season in terrestrial plants. The role of foliar uptake is uncertain. Mosses accumulate lead from the atmosphere and are often used as biological monitors of airborne lead.

Elevated lead levels are also found in terrestrial invertebrates and vertebrates from contaminated areas.

1.2.7 Intake of lead shot

Lead shot taken by birds into their gizzards is a source of severe lead contamination. It results in high organ levels of lead in blood, kidney, liver, and bone.

1.3 Toxicity to Microorganisms

In general, inorganic lead compounds are of lower toxicity to microorganisms than are trialkyl- and tetraalkyllead compounds. Tetraalkyllead becomes toxic by decomposition into the ionic trialkyllead.

One of the most important factors which influence the aquatic toxicity of lead is the free ionic concentration, which affects the availability of lead to organisms. The toxicity of inorganic lead salts is strongly dependent on environmental conditions such as water hardness, pH, and salinity, a fact which has not been adequately considered in most toxicity studies.

There is evidence that tolerant strains exist and that tolerance may develop in others.

1.4 Toxicity to Aquatic Organisms

Lead is unlikely to affect aquatic plants at levels that might be found in the general environment.

In the form of simple salts, lead is acutely toxic to aquatic invertebrates at concentrations above 0.1 and >40 mg/litre for freshwater organisms and above 2.5 and >500 mg/litre for marine organisms. For the same species, the 96-h LC_{50}s for fish vary between 1 and 27 mg/litre in soft water, and between 440 and 540 mg/litre in hard water. The higher values for hard water represent nominal concentrations. Available lead measurements suggest that little of the total lead is in solution in hard water. Lead salts are poorly soluble in water, and the presence of other salts reduces the availability of lead to organisms because of precipitation. Results of toxicity tests should be treated with caution unless dissolved lead is measured.

In communities of aquatic invertebrates, some populations are more sensitive than others and community structure may be adversely affected by lead contamination. However, populations of invertebrates from polluted areas can show more tolerance to lead than those from non-polluted areas. In other aquatic invertebrates, adaptation to hypoxic conditions can be hindered by high lead concentrations.

Young stages of fish are more susceptible to lead than adults or eggs. Typical symptoms of lead toxicity include spinal deformity and blackening of the caudal region. The maximum acceptable toxicant limit (MATC) for inorganic lead has been determined for several species under different conditions and results range from 0.04 mg/litre to 0.198 mg/litre. The acute toxicity of lead is highly dependent on the presence of other ions in solution, and the measurement of dissolved lead in toxicity tests is essential for a realistic result. Organic compounds are more toxic to fish than inorganic lead salts.

There is evidence that frog and toad eggs are sensitive to nominal lead concentrations of less than 1.0 mg/litre in standing water and 0.04 mg/litre in flow-through systems; arrested development and delayed hatching have been observed. For adult frogs, there are no significant effects below 5 mg/litre in aqueous solution, but lead in the diet at 10 mg/kg food has some biochemical effects.

1.5 Toxicity to Terrestrial Organisms

The tendency of inorganic lead to form highly insoluble salts and complexes with various anions, together with its tight binding to soils, drastically reduces its availability to terrestrial plants via the roots. Translocation of the ion in plants is limited and most bound lead stays at root or leaf surfaces. As a result, in most experimental studies on lead toxicity, high lead concentrations in the range of 100 to 1000 mg/kg soil are needed to cause visible toxic effects on photosynthesis, growth, or other parameters. Thus, lead is only likely to affect plants at sites of very high environmental concentrations.

Ingestion of lead-contaminated bacteria and fungi by nematodes leads to impaired reproduction. Woodlice seem unusually tolerant to lead, since prolonged exposure to soil or grass litter containing externally added lead salts had no effect. Caterpillars maintained on a diet containing lead salts show symptoms of toxicity leading to impaired development and reproduction.

The information available is too meagre to quantify the risks to invertebrates during the decomposition of lead-contaminated litter.

Lead salts are only toxic to birds at a high dietary dosage (100 mg/kg or more). Almost all of the experimental work is on chickens and other gallinaceous birds. Exposure of quail from hatching and up to reproductive age resulted in effects on egg production at dietary lead levels of 10 mg/kg. Although a variety of effects at high dosage have been reported, most can be explained as a primary effect on food consumption. Diarrhoea and lack of appetite, leading to anorexia and weight loss, are the primary effects of lead salts. Since there is no experimental evidence to assess effects on other bird species, it is necessary to assume a comparable sensitivity. If this is so, then it is highly improbable that environmental exposure would cause adverse effects.

Metallic lead is not toxic to birds except at very high dosage when administered in the form of powder. It is highly toxic to birds when given as lead shot; ingestion of a single pellet of lead shot can be fatal for some birds. The sensitivity varies between species and is dependent on diet. Since birds have been found in the wild with large numbers of lead shot in the gizzard (20 shot is not unusual), this poses a major hazard to those species feeding on river margins and in fields where many shot have accumulated.

There is little information on the effects of organolead compounds. Trialkyllead compounds produced effects on starlings dosed at 0.2 mg/day; 2 mg/day was invariably fatal.

There are too few reports to draw conclusions about the effects of lead on non-laboratory mammals. Wild rats showed similar effects to their laboratory counterparts.

1.6 Toxic Effects in the Field

Most work on plant tolerance to lead has concentrated on plants growing on mining wastes, naturally highly contaminated areas, and roadside verges. Tolerance has only been found in populations of a few plant species.

No effect on the reproduction of birds nesting near highways has been observed. Toxic effects have been observed in pigeons in urban areas, the kidneys being most frequently affected.

Lead poisoning, due to the ingestion of lead shot, is a cause of death for large numbers of birds. In these cases, lead shot is found in the gizzards, and lead levels are elevated in the liver, kidneys, and bones.

A recurring incident of massive bird kills in estuaries near to industrial plants manufacturing leaded "anti-knock" compounds has been reported. The total lead content of the livers was sufficiently high to cause mortalities: lead was mostly present in the alkyl form.

2. PHYSICAL AND CHEMICAL PROPERTIES

Details of the physical and chemical properties of lead are given in Environmental Health Criteria 3: Lead (WHO, 1977).

Lead (atomic number, 82; atomic weight, 207.19; specific gravity, 11.34) is a bluish or silvery-grey soft metal. The melting point is 327.5 °C and the boiling point, at atmospheric pressure, is 1740 °C. It has four naturally occurring isotopes: 208, 206, 207, and 204, in order of abundance. The isotopic ratios for various mineral sources are sometimes substantially different. This property has been used to carry out non-radioactive tracer environmental and metabolic studies.

Although lead has four electrons in its valence shell, only two ionize readily. The usual oxidation state of lead in inorganic compounds is, therefore, +2 rather than +4. The inorganic compounds of lead are generally poorly soluble, with the exception of the nitrate, the chlorate, and, to a much lesser degree, the chloride. Some of the salts formed with organic acids, e.g., lead oxalate, are also insoluble.

Under appropriate conditions of synthesis, stable compounds are formed in which lead is directly bound to a carbon atom. Tetraethyllead and tetramethyllead are well-known organolead compounds. They are of great importance owing to their extensive use as fuel additives. Both are colourless liquids. Their volatility is lower than for most fuel components. The boiling point of tetramethyllead is 110 °C and that of tetraethyllead is 200 °C. By contrast, the boiling point range for gasoline hydrocarbons is 20 to 200 °C. Evaporation of gasoline tends to concentrate tetraethyllead and tetramethyllead in the liquid residue.

Both tetramethyllead and tetraethyllead decompose at, or somewhat below, the boiling point. Analysis of automobile exhaust gases shows that the ratio of tetramethyllead to tetraethyllead increases as the engine warms up, indicating that tetramethyllead is more thermostable than tetraethyllead. These compounds are also decomposed by ultraviolet light and trace chemicals in air such as halogens, acids, or oxidizing agents.

3. SOURCES OF LEAD IN THE ENVIRONMENT

Details of the sources of lead are given in Environmental Health Criteria 3: Lead (WHO, 1977). The relevant chapter is summarized here.

The major sources of lead in the environment, of significance to living organisms, arise from lead mining and the refining and smelting of lead and other metals. The major dispersive, non-recoverable use of lead is in the manufacture and application of alkyllead fuel additives.

From a mass balance point of view, the transport and distribution of lead from stationary or mobile sources is mainly via air. Although large amounts are probably also discharged into soil and water, lead tends to localize near the points of such discharge. Lead that is discharged into the air over areas of high traffic density falls out mainly within the immediate metropolitan zone. The fraction that remains airborne (about 20%, based on very limited data) is widely dispersed. Residence time for these small particles is of the order of days and is influenced by rainfall. In spite of widespread dispersion, with consequent dilution, there is evidence of lead accumulation at points extremely remote from human activity, for example in glacial strata in Greenland. The concentration of lead in air varies from 2-4 $\mu g/m^3$ in large cities with dense automobile traffic to less than 0.2 $\mu g/m^3$ in most suburban areas, and still less in rural areas.

4. UPTAKE, LOSS, AND ACCUMULATION IN ORGANISMS

Lead is accumulated into many organisms, in many habitats. The following is a selection rather than an exhaustive review. Examples of experimentally determined bioaccumulation factors are given in Tables 1 and 2.

4.1 Controlled Experimental Studies

4.1.1 Model ecosystems

Appraisal

In aquatic and aquatic/terrestrial model ecosystems, uptake by primary producers and consumers seems to be determined by the bioavailability of the lead. Bioavailability is generally much lower whenever organic material, sediment, or mineral particles (e.g., clay) are present. In many organisms, it is unclear whether lead is adsorbed onto the organism or actually taken up. Consumers take up lead from their contaminated food, often to high concentrations but without biomagnification.

Vighi (1981) constructed a simple trophic chain model ecosystem consisting of the alga *Selenastrum capricornutum*, the water flea *Daphnia magna*, and the guppy *Lebistes reticulatus*, and introduced lead as lead nitrate. Concentration factors for the various organisms are given in Table 1. He calculated uptake rates and half-lives for loss of lead. The time taken to reach half the equilibrium concentration of lead in tissues of the organisms ("half-life of uptake") was 5.3 days for the alga, 7.7 days for the water flea, and 25.7 days for total uptake into the fish. Uptake of lead into the guppy was split into two components, that from water and that from food. Half-life of uptake from water was 7.7 days whereas, from food, it was 33 days. Half-lives for loss of lead were calculated as 9 days for fish which had received their lead only from water, and as 40 days for fish which had received lead from food.

Lu et al. (1975) established aquatic/terrestrial model ecosystems based on three different soil types. At the beginning of the experiment, lead chloride was incorporated into the soil. Sorghum seeds were sown in the soil, and algae, daphnids, and pond snails were introduced into the water. On day 7, salt marsh caterpillars were introduced to feed on the sorghum, and, on day 27, mosquito larvae were added to the water. On day 30, some mosquito larvae were removed for analysis and mosquito fish were added to the water to eat the remaining larvae. The experiment was terminated on day 33. Results differed greatly according to the soil type. Using silica sand, with a natural lead concentration of 0.122 mg/kg and 10 mg/kg lead chloride added, the lead levels in organisms were higher than with other soils. With

2

Table 1. Accumulation of lead into aquatic organisms

Organism	Life-stage/ size	Test type[a]	Organ[b]	Temperature (°C)	pH	Compound	Duration (days)	Exposure (µg/litre)	Bioconcentration factor[c]	Reference
Green alga		D	WP	21.1-24.7	7.2-7.8	nitrate	7	4.5	70 000	Vighi (1981)
(*Selenastrum*		D	WP	21.1-24.7	7.2-7.8	nitrate	28	4.5	102 222	
capricornutum)		D	WP	21.1-24.7	7.2-7.8	nitrate	7	40.1	27 431	
		D	WP	21.1-24.7	7.2-7.8	nitrate	28	40.1	32 419	
Pondweed (*Elodea nuttallii nuttallii*)	adult	A	WP	25			30	25	6200	Nakada et al. (1979)
Water hyacinth	adult	A	top	23-27		nitrate	16	1000	492	Muramoto &
(*Eichhornia*	adult	A	roots	23-27		nitrate	16	1000	6200	Oki (1983)
crassipes)	adult	A	leaves	23-27		nitrate	28	1000	5.89	Kay & Haller (1986)
Oyster (*Crassostrea gigas*)		B	WB			chloride	21	100	13.4[d]	Watling (1983a)
Oyster (*Crassostrea margaritacea*)		B	WB			chloride	21	100	17[d]	Watling (1983a)
Marine mollusc (*Perna perna*)		B	WB			chloride	21	100	27.1[d]	Watling (1983a)
Mussel (*Choromytilus meridionalis*)		B	WB			chloride	21	100	31.7[d]	Watling (1983a)
Mussel	6-7 cm	A	kidney	15		nitrate	13	100	3000	Coombs
(*Mytilus edulis*)	6-7 cm	A	kidney	15		citrate	13	100	10 000	(1977)

Table 1 (contd).

Organism	Life-stage/ size	Test type[a]	Organ[b]	Temperature (°C)	pH	Compound	Duration (days)	Exposure (µg/litre)	Bioconcentration factor[c]	Reference
Water flea (*Daphnia magna*)		D	WB	21.1-24.7	7.2-7.8	nitrate	7	4.5	2905	Vighi (1981)
		D	WB	21.1-24.7	7.2-7.8	nitrate	7	315 µg/g	0.04[e]	
		D	WB	21.1-24.7	7.2-7.8	nitrate	28	4.5	5140	
		D	WB	21.1-24.7	7.2-7.8	nitrate	28	460 µg/g	0.05[e]	
		D	WB	21.1-24.7	7.2-7.8	nitrate	7	35.7	756	
		D	WB	21.1-24.7	7.2-7.8	nitrate	7	1100 µg/g	0.025[e]	
		D	WB	21.1-24.7	7.2-7.8	nitrate	28	35.7	1903	
		D	WB	21.1-24.7	7.2-7.8	nitrate	28	1300 µg/g	0.05[e]	
Snail (*Physa integra*)	6-15 mm	C	WB	15	7.1-7.7	nitrate	28	32	3750	Spehar et al. (1978)
Amphipod (*Gammarus pseudolimnaeus*)	5-7 mm	C	WB	15	7.1-7.7	nitrate	28	32	6250	Spehar et al. (1978)
Caddisfly (*Brachycentrus sp.*)	naiad 5-8 mm	C	WB	15	7.1-7.7	nitrate	28	32	8400	Spehar et al. (1978)
Stonefly (*Pteronarcys dorsata*)	naiad 20-40 mm	C	WB	15	7.1-7.7	nitrate	28	32	7800	Spehar et al. (1978)
Stonefly (*Pteronarcys californica*)	naiad	D	WB	3-9	7.0-7.2	nitrate	14	1080	656	Nehring (1976)
Mayfly (*Ephemerella grandis*)	naiad	D	WB	3-9	7.0-7.2	nitrate	14	4900	14 913	Nehring (1976)

Table 1 (contd).

Carp (*Cyprinus carpio*)	10-14 g	A	viscera	14.5-16.5	6.9	nitrate	2	10 000	4200	Muramoto (1980)
	10-14 g	A	gills	14.5-16.5	6.9	nitrate	2	10 000	304	
Pumpkinseed sun-fish (*Lepomis gibbosus*)	10-20 g	A	WB	18-20	6.0	nitrate	8	40	4.88[f]	Merlini & Pozzi (1977)
	12-21 g	A	WB	18-20	7.5	nitrate	8	40	1.86[f]	
Goby (*Gillichthys mirabilis*)	6-38 g	A	spleen	20-25		acetate	8	265	79.4	Somero et al. (1977)
	6-38 g	A	gills	20-25		acetate	8	265	78.5	
	6-38 g	A	fins	20-25		acetate	8	265	78.5	
Guppy (*Lebistes reticulatus*)	150-200 mg	D	WB	21.1-24.7	7.2-7.8	nitrate	7	3.8	654	Vighi (1981)
	150-200 mg	D	WB	21.1-24.7	7.2-7.8	nitrate	7	4.6	1081[g]	
	150-200 mg	D	WB	21.1-24.7	7.2-7.8	nitrate	7	13 µg/g	0.38[e]	
	150-200 mg	D	WB	21.1-24.7	7.2-7.8	nitrate	28	3.8	1072	
	150-200 mg	D	WB	21.1-24.7	7.2-7.8	nitrate	28	4.6	3459[g]	
	150-200 mg	D	WB	21.1-24.7	7.2-7.8	nitrate	28	23 µg/g	0.7[e]	
	150-200 mg	D	WB	21.1-24.7	7.2-7.8	nitrate	7	33.5	197	
	150-200 mg	D	WB	21.1-24.7	7.2-7.8	nitrate	7	35.5	367[g]	
	150-200 mg	D	WB	21.1-24.7	7.2-7.8	nitrate	7	27 µg/g	0.48[e]	
	150-200 mg	D	WB	21.1-24.7	7.2-7.8	nitrate	28	33.5	359	
	150-200 mg	D	WB	21.1-24.7	7.2-7.8	nitrate	28	25.5	1015[g]	
	150-200 mg	D	WB	21.1-24.7	7.2-7.8	nitrate	28	68 µg/g	0.52[e]	
Rainbow trout (*Salmo gairdneri*)	1.0 g	D	WB	14-15.8	7.7-8.1	tetramethyl	7	3.5	725.7[d]	Wong et al. (1981)

[a] A = static conditions (water changed for duration of study); B = water renewed daily; C = flow-through conditions (lead concentration in water continuously maintained).

[b] WB = whole body; WP = whole plant.

[c] Bioconcentration factor = concentration in organism/concentration in medium (calculated on a dry weight basis unless otherwise stated).

[d] Wet weight.

[e] Calculated on lead content of food source; alga for *Daphnia* and *Daphnia* for guppy.

[f] Based on radioactive tracer.

[g] Exposure period from sowing of seed to 30 days post-emergence.

Table 2. Accumulation of lead into terrestrial organisms

Organism	Age	Route	Organ[a]	Compound	Duration (days)	Exposure (mg/kg)	Bioconcentration factor	Reference
Corn (*Zea mays*)		soil	shoots	nitrate	30[b]	4233	0.07	Zimdahl et al. (1978)
		soil	roots	nitrate	30[b]	4233	0.33	
		soil	shoots	sulfate	30[b]	4564	0.05	Zimdahl et al. (1978)
		soil	roots	sulfate	30[b]	4564	0.08	
Sugarbeet (*Beta vulgaris*)		soil	shoots	nitrate	30[b]	4233	0.23	Zimdahl et al. (1978)
		soil	roots	nitrate	30[b]	4233	1.3	
		soil	shoots	sulfate	30[b]	4564	0.04	Zimdahl et al. (1978)
		soil	roots	sulfate	30[b]	4564	0.15	
Bean (red kidney)		soil	shoots	nitrate	30[b]	4233	0.08	Zimdahl et al. (1978)
		soil	roots	nitrate	30[b]	4233	1.0	
		soil	shoots	sulfate	30[b]	4564	0.01	Zimdahl et al. (1978)
		soil	roots	sulfate	30[b]	4564	0.07	
Wheat (*Tritium aestivum*)		soil	shoots	nitrate	30[b]	4233	0.02	Zimdahl et al. (1978)
		soil	roots	nitrate	30[b]	4233	0.2	
		soil	shoots	sulfate	30[b]	4564	0.009	Zimdahl et al. (1978)
		soil	roots	sulfate	30[b]	4564	0.07	
Earthworm (*Eisenia foetida*)		sewage	WB	acetate	35	2500	0.07	Hartenstein et al. (1980)
American kestrel (*Falco sparverius*)	nestling	oral	kidney	metallic	10	25[d]	0.084[c]	Hoffman et al. (1985a)
	nestling	oral	liver	metallic	10	25[d]	0.05[c]	
Starling (*Sturnus vulgaris*)	adult	oral	kidney	triethyl	11	2.85[d]	0.65[c]	Osborn et al. (1983)
	adult	oral	kidney	trimethyl	11	2.85[d]	1.9[c]	

[a] WB = whole body; [b] Exposure period from sowing of seed to 30 days post-emergence.
[c] Wet weight; [d] mg/kg per day.

this soil, lead levels were as follows: water 0.013, algae 275, daphnids 187, snails 334, mosquito larvae 403, fish 13, sorghum leaves 497, and sorghum roots 695 mg/kg. Using silica sand with 10% of silty clay loam (natural lead content 4.5 mg/kg) and 10 mg lead chloride/kg added, uptake into all organisms was markedly less. Lead levels were as follows: water 0.002, algae 114, daphnids 85, snails 56, mosquito larvae 80, fish 1, sorghum leaves 1, and sorghum roots 5 mg/kg. Lead appears to be very strongly bound to even small amounts of fine soil material and, therefore, unavailable to organisms.

4.1.2 Aquatic organisms

Appraisal

The uptake and accumulation of lead by aquatic organisms from water and sediment are influenced by various environmental factors such as temperature, salinity, and pH, as well as humic and alginic acid content.

In contaminated aquatic systems, almost all of the lead is tightly bound to sediment. Only a minor fraction is dissolved in the water, even in the interstitial water.

The lead uptake by fish reaches equilibrium only after a number of weeks of exposure. Lead is accumulated mostly in gill, liver, kidney, and bone.

Fish eggs show increasing lead levels with increased exposure concentration, and there are indications that lead is present on the egg surface but not accumulated in the embryo.

In contrast to inorganic lead compounds, tetraalkyllead is rapidly taken up by fish and rapidly eliminated after the end of the exposure.

Aickin & Dean (1978) exposed 47 bacterial strains and 9 strains of fungi to 300 mg lead/litre (as lead acetate), for 48 h and 7 days, respectively, during the stationary phase of the growth cycle. The uptake of lead was 0.1% to 36% of the dry weight in the bacterial strains and 4% to 19% in the fungi, and was greater than the uptake of copper or cadmium in comparable experiments. When the uptake in 10 bacterial strains was compared, using other, less soluble sources of lead, a general reduction in the amount of lead accumulated was found. Even with lead nitrate, which is soluble, there was reduced uptake in seven out of ten strains. Very little lead was taken up when the metal was added to the medium as lead tetraphenyl. Metallic lead was taken up by many strains to a greater extent than either lead sulfide or the lead oxides.

Four aquatic plant species were exposed by van der Werff & Pruyt (1982) to lead nitrate at concentrations of 1 and 10 μmol/litre for 41 to 46 days and 70 to 73 days. They found that, at both harvest times, the submerged *Elodea nuttallii* and partly-submerged *Callitriche platycarpa* had a higher tissue lead content than the floating species *Spirodela polyrhiza* and *Lemna gibba.* Lead was found in the shoots, roots, and rosettes of *Callitriche* in descending order after 43 days at both concentrations of lead. Roots contained 3 mg lead/kg after exposure to 1 μmol/litre and 13.0 mg/kg after exposure to 10 μmol/litre. Shoots contained nearly 2.5 times more lead than roots, and rosettes less than half as much. Nakada et al. (1979) exposed the submerged plant *Elodea nuttallii* to lead concentrations of 0.025, 0.05, 0.1, and 0.5 mg/litre. After 30 days, the lead content of the plants (with roots removed) was calculated on a dry weight basis. Concentration factors were 6200, 4300, 2800, and 630, respectively. When lead accumulation was monitored in a mixed solution of lead, cadmium, copper, and zinc, the concentration factor was found to be lower than when lead alone was given.

In studies by Kay et al. (1984) the water hyacinth *(Eichhornia crassipes)* was exposed to solutions containing lead nitrate at 0 to 5 mg lead/litre for 6 weeks. The accumulation of lead was dose-related and in the order of roots > stems > leaves. Lead concentrations at similar levels of exposure were only slightly greater after 6 weeks than after 3 weeks. The highest level (5467 mg/kg dry weight) was observed in roots after 6 weeks, following exposure to 5 mg/litre. The results were a compilation of two studies run in the spring and autumn in Florida, USA, lead uptake being consistently higher in the autumn. Kay & Haller (1986) found a concentration factor of 5.89 in water hyacinth leaves at a water concentration of 1 mg/litre. The highest concentration factor was observed at the lowest dose tested; this might indicate that there is a limit on the maximum uptake of the metal by this plant. In further studies, Kay & Haller (1986) exposed water hyacinth to lead nitrate (0 to 5 mg lead/litre) for 4 weeks. Water hyacinth weevils feeding on the leaves, which had been exposed to 5 mg lead/litre, showed concentration factors of 8.89 and 4.5 over the water and the leaves, respectively.

When Meyer et al. (1986) exposed dragonfly larvae to lead nitrate at 20 μg lead/litre for 6 weeks at 15 °C, they found significant accumulation of lead in the fat, midgut, and rectum (0.55, 0.38, and 0.42 mg/kg, respectively). No significant lead residues were found in the brain. The highest levels (1 mg/kg wet weight) were found in the integument. However, this result is not significantly different from controls, which also showed high lead levels (0.8 mg/kg) in the exoskeleton.

In studies by Pringle et al. (1968), mature eastern oysters were exposed to lead in the water at 25, 50, 100, or 200 μg/litre for 49 days. The final concentrations of lead in soft tissues were 17, 35, 75, and 200 mg/kg, respectively. This represents a lead uptake of 0.35, 0.71, 1.50, and 4.00 mg/kg per day for the four exposure levels, respectively.

Coombs (1977) exposed mussels (*Mytilus edulis*) to lead, either as nitrate or complexed with citrate, humic and alginic acids, or pectin, for 13 days at 0.1 mg lead/litre. All tissues showed increasing absorption of the metal over time, but highest concentrations were found in the kidney (Table 1). The uptake rate and total accumulation of lead in all tissues were three to four times higher with lead citrate than with nitrate. The other complexes were not so effective in increasing lead uptake; at best they produced 1.5- to 2-fold increases.

When Anderson (1978) exposed crayfish *(Orconectes virilis)* to lead acetate at concentrations of 0, 0.5, 1, and 2 mg lead/litre, he found, over the 40-day exposure period, a marked increase in the lead content of both gills and exoskeleton as water concentration and exposure time increased. There was also an increase in the lead concentration in muscle and viscera, but this was not significantly affected by either treatment concentration or length of exposure.

Ray et al. (1981) exposed three species of marine invertebrates, *Nereis virens, Crangon septemspinosa,* and *Macoma balthica,* to two sediments which contained different amounts of lead. The sediments had no added lead but were collected from different areas; they also contained different amounts of other metals (copper, zinc, and cadmium). Sediment A (48% sand; lead at 96.2 mg/kg dry weight) contained lower levels of all metals than did sediment B (33% sand; lead at 243.9 mg/kg dry weight). Animals were exposed to the sediment for 30 days. Although *N. virens* showed no increase in lead tissue concentrations in sediment A, other species in sediment A, and all species in sediment B, revealed tissue lead increases over time. Concentration factors ranged from 0.01 to 0.06, higher tissues levels being attained after exposure to sediment B.

In a similar study, Lewis & McIntosh (1986) exposed the freshwater isopod *Asellus communis* to two contaminated sediments in water at three different pH levels for 20 days. Sediment A, a clay loam, contained higher metal levels (lead at 367 mg/kg dry weight) than sediment B, a silt loam (lead at 266 mg/kg dry weight). Higher lead levels were found in the corresponding interstitial water in sediment A (lead at 10.2 μg/litre and 5.1 μg/litre for A and B sediments, respectively). Lead accumulation from sediment was significant in sediment A at pH 4.5 and 5.5 but not at pH 7.5, and from water, only at pH 4.5. After 20 days exposure to sediment A at pH 4.5, concentration factors were 1.4, in terms of sediment, and 39 000, in terms of water, corresponding to lead levels in *Asellus* of 510 mg/kg dry weight. There was no significant accumulation from sediment B.

Maddock & Taylor (1980) investigated the uptake of organolead compounds by shrimp, mussel, and dab (a flatfish) in short-term experiments and in mussel and dab in long-term experiments. For the short-term exposure, shrimps (*Crangon crangon*), mussels *(Mytillus edulis*), and dabs (*Limanda limanda*) were held in concentrations of teramethyl-, tetraethyl-, trimethyl-, and triethyllead up to the level of the 96-h LC_{50}. This experiment measured the lead content of animals used in the tests to determine acute toxicity. Results should, therefore, be treated with caution because of some mortality at the higher end of the range. Bioconcentration factors were higher for tetraalkyllead than for trialkyllead; they lay between 20 and 650 for the two tetraalkyl compounds in the three species, and between 1 and 24 for the two trialkyl compounds in the same three species. Mussels exposed to either 0.01, 0.05, or 0.10 mg trimethyllead chloride/litre (96-h LC_{50} = 0.5 mg/litre) showed maximum uptake of lead within 9 days, and further exposure over 35 days failed to cause any further tissue accumulation of lead. Uptake was dose-related; the mean tissue content after exposure at 0.10 mg/litre for 21 days was 68 mg/kg wet weight, representing a bioconcentration factor of 90. The greatest tissue concentration occurred in the gill with the digestive gland, gonad, and foot containing progressively less lead. Loss of lead was rapid when the animals were transferred to clean water, with a mean half-time of 3 to 4 days. Results for triethyllead chloride uptake and loss by mussels were very similar. The authors conducted a comparison between uptake of organic and inorganic lead in mussels; it is clear that inorganic lead is accumulated to a much greater extent. Dabs were exposed to either 1.0 or 2.0 mg trimethyllead/litre (96-h LC_{50} = 24.6 mg/litre), or to either 0.1 or 0.2 mg triethyllead/litre (96-h LC_{50} = 1.17 mg/litre) for 41 days. With the exception of liver uptake of trimethyllead, where equilibrium was reached after 20-days, uptake into liver and muscle was linear over this period. Uptake by liver and muscle was similar, with average tissue lead levels at around 30 mg/kg wet weight; this represented bioconcentration factors of 2 for trimethyl- and 12 for triethyllead. Loss was slow with half-times in excess of 41 days where these could be determined.

In studies by Holcombe et al. (1976), brook trout *(Salvelinus fontinalis)* were exposed to lead nitrate concentrations of 0.9 to 474 µg lead/litre for three generations over a period of 3 years. Gill, liver, and kidney tissues of first and second generation trout accumulated the greatest amount of lead. In the first generation fish, these organs appear to reach equilibrium after 20 weeks exposure to 235 and 474 µg lead/litre, but not at lower concentrations. A equilibrium of lead residues was reached in liver and kidney tissue from second generation fish after 70 weeks of exposure to 119 µg lead/litre. Lead residues in gill tissue continued to increase throughout the 100 weeks of the first and second generations. In the third generation, samples of eggs at spawning, and alevins, 4 weeks after hatch, showed that lead residues increased with higher exposure concentrations. Although eggs showed increasing lead levels with increased exposure concentrations, newly hatched alevins had negligible residues. This indicates that the lead was present in the egg membrane but not accumulated by the embryo. Juvenile alevins accumulated lead up to an age of 8 weeks and then showed a reduced concentration of lead after 12 weeks. It is not clear from the results whether this represents lead loss or simply a reduced rate of uptake in the larger fish.

Merlini & Pozzi (1977) exposed the pumpkinseed sunfish to lead nitrate (traced with ^{203}Pb) at 40 µg lead/litre for up to 8 days at pH 6.0 and 7.5. The fish accumulated nearly three times as much lead from water at the lower pH (Table 1).

In studies by Hodson et al. (1978b), 4-month-old rainbow trout were exposed to nominal concentrations of lead between 0 and 1000 μg/litre (at pHs of 6, 8, and 10 for 3 days, and 7, 8, and 9 for 2 days). It was found that blood lead levels increased as the pH of the test water decreased from 10 to 6. The highest blood lead level (approximately 10 000 μg/litre) was recorded after exposure at pH 6 and a water concentration of 180 μg lead/litre. This represents a concentration factor of about 50 in blood over water. The authors calculated that a decrease by a pH unit of 1.0, from any reference pH, resulted in an increase of blood lead by a factor of 2.1. Blood lead was found to be in equilibrium with lead in the water within 48 h of exposure.

Somero et al. (1977) exposed the estuarine teleost *Gallichthys mirabilis* to lead acetate concentrations of 2650 mg lead/litre for 36 days in 100% sea water (3.36% salinity) and in 75%, 50%, and 25% sea water. The lead content of all tissues studied showed an increase with decreasing salinity. Highest levels were in the spleen, ranging from a concentration factor (on a dry weight basis) of 74.4, for 100% sea water, to 137.7, for 25% sea water. The same authors also exposed the fish to two different temperature regimes, 10 °C and 20-25 °C, for 42 days in normal sea water. They found that a higher temperature resulted in a higher tissue lead content.

Muramoto (1980) held carp *(Cyprinus carpio)* for 48 h in lead nitrate concentrations of between 0 and 20 mg lead/litre, with and without one of the three complexans, EDTA, NTA, or DTPA. The accumulation of lead in both viscera and gills was dose-related, with the highest levels for viscera and gills being 86 000 and 4560 mg/kg dry weight, respectively. The complexans reduced the uptake of lead at all dose levels. Concentrations in viscera ranged from 399 to 690 mg/kg for the three complexans, and 298 to 645 mg/kg in gills, after exposure to 20 mg/litre lead (which had given the above levels without chelating agents). It is not clear whether the levels of lead in the gills represented uptake into the tissue or adsorption onto the exterior surfaces.

Wong et al. (1981) exposed rainbow trout to tetramethyllead at 24 μg/litre for up to 10 days. Because of the high volatility and low water solubility of the compound, the authors designed an apparatus specifically for the test. The water was changed completely every 2 h in a flow-through system to which the tetramethyllead was continuously added. They found that most of the alkyllead was accumulated in the intestinal lipid (concentrations of 63 to 140 mg/kg wet weight), followed, in decreasing order, by gills, skin/head, and air bladder. They also calculated uptake rates and depuration rates for tetramethyllead in rainbow trout. The uptake rate was greatest (1 μg/g of fish/day) at the beginning of exposure, and reached equilibrium by day 7. When exposure stopped and the fish were returned to clean water, levels of tetramethyllead in the tissues decreased rapidly over 3 days and then declined more slowly. Concentrations of alkyllead in tissues had returned to pre-exposure levels within 1 week. Rate constants for loss were 0.58/day for intestinal lipid and 0.29/day for skin and head.

In studies by Ireland (1977), toads *(Xenopus laevis)* were fed with live earthworms containing 10, 308, or 816 mg lead/kg for 4 or 8 weeks. Toads fed the diet containing 10 mg/kg for 8 weeks had significantly less lead in kidney and liver than toads fed 308 mg/kg diet for 4 weeks (or 308 mg/kg for 4 weeks, followed by 816 mg/kg for 4 weeks). Bone and skin lead levels were significantly less after 4 weeks on 10 mg/kg than after 4 weeks on 308 mg/kg diet. No other significant difference was observed. Muscle lead levels did not vary significantly between treatments. Individual organ analysis, within groups, showed high lead levels in kidney, bone, and liver, but low values in skin and muscle. The highest levels were found in kidney and were 19.1, 73.3, and 81.3 mg/kg dry weight at the three dose levels, respectively.

4.1.3 Terrestrial organisms

Appraisal

In bacteria, the majority of lead is associated with the cell wall. A similar phenomenon is also noted in higher plants. Some lead that passes into the plant root cell can be combined with new cell wall material and subsequently removed from the cytoplasm to the cell wall. Of the lead remaining in the root cell, there is evidence of very little translocation to other parts of the plant because the concentration of lead in shoot and leaf tissue is usually much lower than in root. Foliar uptake of lead occurs, but only to a very limited extent.

In animals, there is a positive correlation between tissue and dietary lead concentrations, although tissue concentrations are almost always lower. The distribution of lead within animals is closely associated with calcium metabolism.

Lead shot is typically trapped in the gizzard of birds where it is slowly ground down resulting in the release of lead.

The tetravalent organic form of lead is generally more toxic than the divalent, inorganic form, and its distribution in organisms may not specifically follow calcium metabolism.

When Tornabene & Edwards (1972) incubated two species of bacteria, *Micrococcus luteus* and *Azotobacter* sp., in a medium with a suspended dialysis bag containing lead bromide, the two species took up 490 and 310 mg lead/g whole cells (dry weight), respectively. The authors analysed subcellular fractions of the bacteria and found 99.3% and 99.1%, for the two bacteria, respectively, in the cell wall plus membrane fraction. The remainder of the lead was found in the cytoplasm. The same authors, Tornabene & Edwards (1973), located electron-dense inclusions in cell membranes of *Micrococcus.* Tornabene & Peterson (1975) showed that the lead was not specifically bound to lipid fractions in the cell membrane but that the membrane provided a suitable substrate in which aggregations of lead could form.

Zimdahl et al. (1978) sowed maize *(Zea mays)*, sugarbeet, bean, and wheat in soil dosed with lead nitrate or sulfate (0 to 5000 mg lead/kg). Lead uptake into shoots and roots (on a dry weight basis) was measured 30 days after emergence. It was found that more lead was taken up into the roots than into the shoots (Table 2). Although the data are not conclusive, the authors suggest that less lead is taken up when soil is treated with lead sulfate than with lead nitrate. In a 2-year study, Baumhardt & Welch (1972) grew *Zea mays* in the field where lead acetate had been applied to the soil at rates of 0 to 3200 kg lead/ha. The lead contents of the plants for the 0 and 3200 kg/ha treatments were, respectively, 2.4 and 37.8 mg/kg for young whole plants, 3.6 and 27.6 mg/kg for leaves at tasselling, and 4.2 and 20.4 mg/kg for whole plants at grain harvest. The lead content of grain was unaffected by any of the applications.

Lane & Martin (1977) investigated the uptake of lead into the seed and seedlings of the radish, the location of the lead being determined histochemically. The intact testa prevented uptake of lead into the embryo, but when the testa ruptured during germination, the radicle took up lead readily, as did the rest of the tissues (endosperm). As the seedling developed, lead was concentrated in the radicle and the hypocotyl, with relatively little being transported to the shoot.

In studies by Malone et al. (1974), *Zea mays* was exposed to lead, in either a hydroponic solution or in distilled water, in four different forms: citrate, chloride, nitrate, or EDTA chelate. Lead concentrations in the solutions ranged from 10 to 1000 mg/litre. The uptake of lead was followed using phase-contrast light and electron microscopy. Roots generally accumulated a surface precipitate of lead salts as fairly large crystals. Lead was slowly absorbed into the roots and appeared as much smaller crystals associated primarily with the cell walls. The lead was taken up by dictyosome vesicles which migrated towards the cell wall and ultimately formed extensions of the cell wall. These vesicles fused together to encase the lead crystals within the cell wall material. In some cases, these inclusions projected into the cell cytoplasm. Similar deposits were found in shoots and leaves as the lead was slowly transported throughout the plant. Lead was never associated with the phloem or its companion cells and never with the guard cells of the epidermal stomata. Thus lead was excluded from the biochemically active plasmalemma.

Hemphill & Rule (1975) applied solutions of radioactively labelled lead nitrate to the leaves of lettuce and radish for a period of 25 days and then grew the plants on for a further 25 days. The lead content and distribution were assessed using scintillation counting and autoradiography. There was some absorption of lead into the leaves and some subsequent translocation, but this was very small. The percentage translocation of applied lead (expressed in terms of the total lead applied) was not more than 0.2%, and generally much less than this, except where contamination with the applied material had possibly occurred. It is not clear whether the total recovery of the labelled lead was estimated.

Dollard (1986) conducted a similar experiment with glasshouse-grown radish, carrot, and French bean plants. In radish, a small amount (0.05% to 0.28%) of the applied lead was transported to the swollen root. This movement occurred through intact or damaged cuticle, and there was some indication that damage to the leaf surface enhanced lead uptake. Carrot plants absorbed 0.43% of foliar-applied lead, but transported it no further than the leaf petiole over the 8- to 12-week period of the experiment. The transport of lead to the tap root was <0.01% of that applied. For the French bean, no movement of lead into pod or seed was detected. The author estimated that up to 35% of root lead in radish could be accounted for by foliar absorption, whereas in carrot this would be no more than 3% (based on lead deposition rates from the atmosphere close to roads).

Beyer et al. (1982) monitored the uptake of metals into earthworms from soil treated with sewage sludge. In all treatments, the concentration of lead in the earthworms correlated with the concentration in soil. There was, however, no bioconcentration of lead into worms, concentration factors being consistently less than 1.0 for soil lead levels ranging between 16 and 43 mg/kg.

In studies by Straalen & Meerendonk (1987), adult collembola *(Orchesella cincta)*, collected from an unpolluted pine forest and cultured in the laboratory, were fed with green algae on paper disks. Lead nitrate solution was added to the food suspension. The concentration of lead in the food ranged from 1600 to 2200 mg/kg dry weight. The study lasted for 8 weeks, contaminated food being fed for the first 4 weeks and clean food for the second 4 weeks. Lead concentrations in the collembola fluctuated within wide limits during the accumulation phase. An average steady state was achieved after approximately 4 weeks, with lead concentrations of approximately 0.2 mg/kg dry weight. This value was obtained for worms with the gut contents cleared. The authors identified three components to the body lead content: gut contents, a "fast body burden", and a "slow body burden". The fast component appeared to be lost during moults. Calculated half-times for loss of lead from the three components were as follows: 0.34 days for gut content, 7.37 days for "fast body burden" and 21.66 days for "slow body burden".

Irwin & Karstad (1972) exposed adult mallard drakes to 17.8, 89, or 178 g of particulate lead per m^2 in a simulated marsh environment for 14 weeks. Lead shot (no. 5) were scattered over the penned area which simulated a marsh area of puddled mud. The number of shot actually ingested per bird is not clear. Lead levels in muscle, liver, and bone increased with increasing exposure. Liver and bone contained higher concentrations; after 14 days exposure to 178 g/m^2, lead concentrations in liver and bone were 28.4 mg/kg wet weight and 176 mg/kg dry weight, respectively.

When Clemens et al. (1975) dosed adult mallard with five lead shot (no. 6) and monitored tissue concentrations of lead over a period of 20 days, they found higher lead tissues levels in birds on a high-fibre diet (12.5% fibre) than on a low-fibre (3%) diet. The highest levels were found in the bone after 16 days (570 mg/kg dry weight) and in the kidney after 12 days (225 mg/kg wet weight), both on the high-fibre diet. In the birds on a low-fibre diet, lead levels peaked in all tissues after 2 to 4 days and then declined. In birds on high-fibre diets, the same was true only for blood. Lead levels did not peak until 12 days in liver, kidney, leg muscle, and bone. In both groups, the pectoral muscle, after an initial rise, showed fluctuating levels with no consistent pattern.

Finley et al. (1976) dosed male and female mallard with either one (no. 4) lead shot or one (no. 4) lead/iron combination shot (with 47% lead). The birds were observed for 4 weeks. The lead levels in liver, kidney, blood, and bone were twice as high in birds dosed with lead alone, reflecting the relative amounts of the metal consumed. Females had double the lead levels of males, except in bone, where the difference was a factor of ten. The levels in females dosed with lead shot were 1.15, 3.53, 0.71, and 112.27 mg/kg for liver, kidney, blood, and bone, respectively. Similar trends were found in eggs laid during the period, with the birds dosed with lead shot laying down more lead in the eggs. The egg contents and shell contained 0.5 and 2.8 mg lead/kg, respectively, after dosing with lead shot.

When mallard were dosed with one lead shot (no. 4), the pre-dosing blood lead level was 83 μg/litre and rose to 317 μg/litre 1 month after dosing. Four weeks after male and female mallard were dosed similarly, lead accumulation was significantly greater in bones with a high medullary content (femur and sternum) than in bones with a lower content (ulna/radius and wing bones). Females always contained higher bone residues than males. The femurs of laying females averaged 488.8 mg lead/kg dry weight compared with 113.6 mg/kg in non-laying females and 9.4 mg/kg in males. When birds were dosed with a second lead shot and analysed 4 weeks later, levels in laying females were unchanged but levels in males had risen by a factor of three. The authors suggested that a saturation level had been reached in the females (Dieter & Finley, 1978; Finley & Dieter, 1978).

Buggiani & Rindi (1980) dosed adult domestic ducks with 24 lead shot (no. 6) once a week for 5 weeks. A second group were dosed for 6 weeks with the same number of shot plus EDTA (1 mmol/kg body weight). At the end of the experiment, lead concentrations were measured in the blood and the nasal glands. Blood lead was three times higher than control levels in both groups. The ratio of nasal gland lead to blood lead was 1 for birds from both groups. Immediately after treatment with lead shot, this ratio was 3 suggesting that the nasal gland is a source of lead excretion in ducks.

In studies by Osborn et al. (1983), starlings *(Sturnus vulgaris)* were orally dosed with solutions of triethyllead or trimethyllead chlorides at concentrations of 0, 200, and 2000 μg/litre per day for 11 days, or until death. All the birds in the low-dose group survived for the full 11 days; birds dosed with trimethyllead accumulated more lead in the brain, kidney, and liver than did triethyllead-treated birds. The highest lead levels were found in the kidney: triethyllead-treated birds contained 1.85 mg/kg wet weight and trimethyllead-treated birds contained 5.38 mg/kg wet weight in their kidneys. Birds given the high dose all died within 6 days, and had higher lead levels in all tissues than birds given the lower dose. In these dead birds, highest lead concentrations were found in the liver of triethyllead-treated birds (40.2 mg/kg wet weight) and the liver and kidney of trimethyl-lead-treated birds, (32.4 and 30.2 mg/kg wet weight, respectively). Osborn (1979) pointed out that metal levels in different tissues of birds should be treated with caution since they depend on many different factors. In particular, levels in dead or dying birds are not comparable to those in healthy birds because of redistribution prior to death. Also, it is not possible to compare exposure of birds in the field with those in the laboratory simply by measuring tissues levels.

4.2 Accumulation in the Field

4.2.1 General considerations

Appraisal

Organisms have been found to incorporate lead from the environment, generally in proportion to the degree of contamination. Lead deposition in a region depends on the air concentrations of the metal, which decrease with the distance from the source.

In shellfish, lead concentrations are higher in the calcium-rich shell than in the soft tissue; they relate to the concentrations in sediment.

Lead concentrations in some marine fish are higher in gills and skin than in other tissues, but this may be largely due to adsorption. Liver levels increase significantly with age.

In dolphins, lead is transferred from mothers to offspring during fetal development and lactation. This might be related to the calcium metabolism.

Ayling (1974) sampled the oyster *Crassostrea gigas* from the Tamar River in Tasmania and found mean dry weight lead concentrations in oysters and mud samples of 0 to 135 mg/kg and 4 to 1500 mg/kg, respectively. The author stated that lead was not taken up through any physiological demand, but was randomly incorporated at the sites containing high concentrations in the mud. When analysing the bivalve *Elliptio complanata* from the Great Lakes for lead levels, Dermott & Lum (1986) found higher levels (10.2 to 25.2 mg/kg) in the shell than in soft tissues (ND to 2.2 mg/kg). Lead was significantly higher in the outer periostracum of the shell than in the inorganic prismatic layer. In spite of high levels at one site contaminated by effluent, lead was not deposited in the prismatic shell layer. Sediment levels in the sampling areas ranged from 29 to 103.3 mg lead/kg. Pringle et al. (1968) found low levels of lead (<0.2 mg/kg in soft tissues) in estuarine molluscs. There was no seasonal variation in lead concentrations.

Enk & Mathis (1977) detected lead in all components of a stream with no industrial contamination. The levels were as follows: water (<0.5 mg/litre), fish (2.47 to 2.88 mg/kg), sediment (8.3 mg/kg), aquatic insects (6.83 to 12.59 mg/kg), snails (13.64 mg/kg).

When Gilmartin & Revelante (1975) analysed anchovy and sardine from the Adriatic Sea, the highest lead concentrations were found in the gills (6.8 and 6.5 mg/kg wet weight, respectively) and skin (4.5 and 4.3 mg/kg wet weight, respectively). Higher liver lead concentrations in anchovy occurred later in the year. For most of the period of study, lead was not detectable in the muscle, digestive system, or liver. Perttila et al. (1982) found that lead increased significantly with age in the Baltic herring *(Clupea harengus)*.

Van Hook (1974) calculated concentration factors for lead into earthworms sampled from the field. Factors were below 1 (range 0.11 to 0.3) for soil lead levels ranging between 15 and 50 mg/kg dry weight. Bagley & Locke (1967) analysed wild birds of many species, from the eastern USA, for tissue lead levels. The majority of birds examined were water-fowl. All the birds were healthy and contained no lead shot. Mean liver residues of lead ranged from 0.5 to 3.7 mg/kg wet weight and mean tibia residues from 2.0 to 13.0 mg/kg wet weight. Martin (1972) and Martin & Nickerson (1973) analysed starlings in the USA for lead and found residues ranging from 0.4 to 13.3 mg/kg in 1970 and 0.12 to 6.6 in 1971.

In studies on the common porpoise *(Phocoena phocoena)* from the east coast of Scotland, Falconer et al. (1983) found that lead residues were below detectable limits (0.5 mg/kg). The sampled animals had died after becoming entangled in cod nets. The tissues analysed were the brain, liver, kidney, heart, and spleen. Honda et al. (1986) sampled striped dolphin *(Stenella coeruleoalba)* and found significant accumulation of lead in the bone of offspring during the suckling period. Significantly more lead was found in adult males than females. The authors suggested that lead was removed from the mother via the milk and as the result of parturition. Lead levels ranged between 0.09 and 0.74 mg/kg wet weight.

4.2.2 Highways and urban areas

Appraisal

Lead concentrations are highest in soils and organisms close to roads where traffic density is high. The lead measured is inorganic and derives almost exclusively from alkyllead compounds added to petrol.

The lead in the soil and in vegetation decreases exponentially with the distance from the road. Lead is also found in the sediments of streams in the vicinity of highways.

Lead contamination increases lead levels in plants and animals in areas close to roads. These levels are positively correlated with traffic volume and proximity of roads.

Most lead deposited is found within 500 m of the road and within the upper few centimetres of soil. It can be assumed that lead levels in soil and biota are not influenced by traffic at distances from roads greater than this.

There is extensive documentation on the occurrence of lead in soil and organisms close to roads.

Khalid et al. (1981) analysed soil samples from different areas of Baghdad, and found that mean levels ranged from 36 mg/kg for an industrial area to 308 mg/kg for north-east Baghdad. It was also found that lead concentrations were highest in areas of high traffic volume and the city centre had higher levels than other areas.

Chow (1970) established, by isotopic composition, that lead detected in soil and dried grass derived exclusively from alkyllead compounds added to petrol. Wheeler & Rolfe (1979) established a double exponential relationship between lead levels in vegetation and distance from the road. The two exponents were assumed to represent particles of different size. Larger particles were deposited within about 5 m of the edge of the road surface. Smaller particles settled more slowly and were deposited within 100 m of the road, though beyond 50 m from the road surface there was little more than a background level of lead. Lead in the smaller particles was more soluble than in the larger. Lead levels were very high close to the road. At a traffic density of 8100 vehicles/day, lead concentrations of 1225 mg/kg soil and 196 mg/kg vegetation were found within 0.3 m of the road. This declined rapidly; soil levels were 526 mg/kg at 1 m, 93 mg/kg at 5 m, and 55 mg/kg at 10 m from the road, with similar falls in vegetation levels. Although the soil had a high capacity to adsorb lead, an estimated 72-76% of the total lead deposited had been lost from the soil by leaching or run-off.

In a similar study (Ward et al., 1975) of a road in New Zealand with a traffic density of 1200 vehicles/day, a similar distribution of lead was noted. All lead deposited could be found within 100 m of the road and within the upper 5 cm of the soil. The authors calculated that the total deposition of lead since the introduction of leaded petrol was around 250 g/metre of road length. Of this, 140 g lead could be accounted for within 250 m of the road side and in the upper 6 cm of soil. Cannon & Bowles (1962) reported that lead levels depend on the traffic volume on the roads and rise to 3000 mg/kg in grass near major road intersections. Lead is also found in streams close to major roads.

Van Hassel et al. (1979) reported little or no difference between the lead concentration in water of roadside streams and that of streams away from highways. There was, however, a significant increase in the lead content of the stream sediment, to which lead is readily adsorbed. Similar results were found by Mudre & Ney (1986) who investigated the lead content of the sediment in a series of tributary streams running into the Chickahominy River in Virginia. The same highway crossed all streams. Levels of lead close to the road were significantly higher than in upstream samples from all streams. Samples taken some distance downstream did not differ from upstream ones; lead contamination was very localized. There were marked differences between streams due to various factors including drain-off from vegetation into the stream, weather, stream flow rates, and traffic density at different times of year.

Ash & Lee (1980) monitored lead in earthworms (two species) from sites close to roads and from low-traffic areas in the United Kingdom. The earthworms were purged of gut contents before analysis, and all results are expressed in terms of dry weight. The control site in rural Scotland showed lead levels of 0.96 and 0.31 mg/kg dry weight in the two species. Close to two major roads, levels were 130 and 341 mg/kg (for the A660 road) and 274 and 500 mg/kg (for the A1 road with greater traffic density) for the two earthworm species, respectively. A city recreational area gave levels of 32 and 76 mg/kg and a site on farmland (300 m from the main A1 road) gave levels of 38 and 26 mg/kg for the two earthworm species, respectively. Goldsmith & Scanlon (1977) measured lead concentrations (excluding gut contents) in earthworms at 6, 12, and 18 m from two roads in Virginia, USA. The roads had traffic volumes of 21 040 and 1085 vehicles/day, respectively. Lead levels in earthworms were 51, 50, and 32 mg/kg dry weight at 6, 12, and 18 m, respectively, from the busier road. At 12 and 18 m from the less busy road, levels were 8.5 and 11.65 mg/kg, respectively.

Price et al. (1974) found that sap-sucking, phytophagous, and insectivorous insects contained, on average, 10.3, 15.5, and 25.0 mg lead/kg, respectively, close to a road. In low-lead areas, the three types of insect showed 4.7, 3.4, and 3.3 mg/kg, respectively. The authors claim evidence for the concentration of lead through food-chains. Giles et al. (1973) came to similar conclusions while measuring lead in phytophagous and carnivorous insects. Beyer & Moore (1980) reported that caterpillars feeding on black cherry leaves contained 76% as much lead as did their food. More lead was found in the insects close to the road than in those further away. Beyer (1986) has questioned the bioconcentration of lead in road-side food-chains, since no study has exhaustively monitored lead in prey and predators of a recognized food-chain. Other explanations of the available data are probable; different species of insects, both prey and predator, have been shown to take up lead to very different degrees.

May & McKinney (1981) showed that lead concentrations in Hawaiian fish, sampled from streams close to roads, ranged from 0.8 to 4.93 mg/kg wet weight in whole fish, with high levels corresponding to high-traffic density. The species sampled included some bottom-feeders, but were mainly fish of the open water. Ney & Van Hassel (1983) measured the whole body lead content of six fish species sampled from a stream flowing under a major highway. Fish were sampled close to the bridge. The residues (means for species) ranged between 7.2 and

19.5 mg/kg dry weight, and species which live in the open water had lower levels than bottom-feeding species. Levels of lead in sediments, benthic invertebrates, and fish were higher at this site than upstream or downstream of the road crossing, indicating localized binding of the metal (Van Hassel et al., 1979, 1980).

Birdsall et al. (1986) measured lead concentrations in sediment and in the tadpoles of bullfrogs *(Rana catesbeiana)* and green frogs *(Rana clamitans)* taken from drains beside roads with different daily traffic volumes and from ponds at least 0.4 km from the nearest road. Sediment samples showed lead concentrations ranging from 7.8 to 40 mg/kg dry weight for ponds and 18 to 940 mg/kg dry weight for highway drains. These were usually 4 to 5 times greater than corresponding levels in tadpoles. Levels in bullfrog tadpoles were 2.6 to 6.0 mg/kg and 0.7 to 270 mg/kg for the ponds and drains, respectively. Green frog tadpoles contained 0.9 to 8.9 mg/kg in ponds and 4.8 to 240 mg/kg in drains. There was a positive correlation between traffic volume and the lead content of sediment and amphibians.

Ohi et al. (1974) determined lead levels in blood, femurs, and kidneys of adult pigeons sampled from rural and urban sites in Japan. Lead levels were highest in femurs, with means ranging from 16.5 to 31.6 mg/kg wet weight over three urban sites, while two rural sites showed mean levels of 2.0 and 3.2 mg/kg. Blood levels showed a similar trend; the urban sites gave 0.15, 0.33, and 0.33 mg/litre while the rural sites showed 0.054 and 0.029 mg/litre. Kidney levels were lower, and also showed a reduced lead level in rural areas. Hutton & Goodman (1980) obtained similar results in pigeons in London, with differences between central London, suburban London, and surrounding rural areas. Getz et al. (1977) sampled four species of song birds from an urban site and rural sites in Illinois, USA. The rural sites were chosen to be at least 2 km from the nearest town and 50 m from any road. Highly significant differences in lead content between urban and rural values were found for all species and in all tissues (feathers, gut, liver, kidney, and femur), except for lung and pectoral muscle, which showed low lead content. Kidney levels in urban areas were 33.9, 98.5, 13.5, and 25.0 mg/kg dry weight for house sparrow, starling, grackle, and American robin, respectively, and in rural areas were 3.5, 3.6, 3.5, and 7.3 mg/kg for the same species, respectively. Grue et al. (1986) found lead levels 3 to 13 times higher in starlings (nestling and adults) breeding near roads than in birds sampled from control sites. There was a less pronounced, but still significant, difference between similarly sited breeding colonies of swallows (Grue et al., 1984).

Jefferies & French (1972) measured the lead in the liver and whole body of 101 small mammals of three species, *Microtus agrestis*, *Clethrionomys glareolus*, and *Apodemus sylvaticus*, sampled from fields or from roadside verges. The mean lead concentration of whole bodies increased from 4.19 mg/kg dry weight for mammals trapped on woodland or arable sites to 5.98 mg/kg on the verges of minor roads and 7.0 mg/kg on the verges of a major road. Vegetation from the same sites averaged 33.4, 42.5, and 306.7 mg/kg dry weight, respectively. Goldsmith &

Scanlon (1977) trapped small mammals in three study areas of roadside verges with different traffic densities. Significantly greater lead levels were found in heavy traffic areas in individuals of three species: *Cryptotis parva*, *Microtus pennsylvanicus*, and *Peromyscus leucopus*. However, no significant difference between areas was found in the shrew *Blarina brevicauda*. These species represent herbivores and carnivores, with the shrews eating predominantly insect prey. Carnivores had higher levels of lead than herbivores. Welch & Dick (1975) found that lead levels in liver, kidney, and bone (but not those in brain, lung, stomach, or muscle) of deer mice *(Peromyscus maniculatus)* were related to the proximity to the road and to traffic volume.

Quarles et al. (1974) found that the lead content of small mammals increased with proximity to the road. In comparable areas, there was 22.7 mg/kg in the shrew *Blarina*, 16.3 mg/kg in the vole *Microtus* and 6.8 mg/kg in the mouse *Peromyscus*. The authors compared published information on the food consumption, food choice, and habits of the three species. The size of the home range was suggested as a contributing factor to differing lead concentrations; the mouse has a much more extensive range than the shrew or vole. Food type, with the insectivorous vole taking in most lead, was also likely to be important. Williamson & Evans (1972) analysed the lead content of a wide variety of invertebrates from roadside verges and also of small mammals which eat these invertebrates. They found no evidence to indicate concentration of lead in food-chains. Although the insectivorous shrews had higher lead levels than their herbivorous neighbours, the shrews contained less lead per unit weight than did their prey.

4.2.3 Industrial sources

Appraisal

Terrestrial and aquatic plants accumulate lead in industrially contaminated environments. In aquatic plant species, lead uptake can occur from both water and sediment, although uptake from sediment usually predominates. Lead levels decrease with distance from the source and are lowest during the active growing season in terrestrial plants. The role of foliar uptake is uncertain. Mosses accumulate lead from the atmosphere and are often used as biological monitors of airborne lead.

Elevated lead levels are also found in terrestrial invertebrates and vertebrates from contaminated areas.

Rains (1971) analysed the lead content of wild oats *(Avena fatua)* growing in the vicinity of a smelter. The area had been subject for more than 70 years to lead contamination from the smelter, which was still in operation. The lead content of the plants increased throughout the year, the lowest levels occurring during the active growing season. Lead levels continued to rise after the ears were fully formed and the upper portions of the plant were dry, and peaked at 500 mg/kg dry weight. Some lead would be taken up from the soil, but the predominant source of the lead would be atmospheric.

Mayes et al. (1977) measured lead uptake into a submerged aquatic plant *Elodea canadensis* in two lakes. The control lake was far from any industrial sources of metal, while the second lake received waste water from an electroplating plant. Specimens of *Elodea* were anchored in each lake into contaminated and non-contaminated sediments. Plants grown in the same water, but in sediment from different sources, had significantly greater lead content when grown on the contaminated sediment. Similarly, *Elodea* accumulated more lead when grown in contaminated water, irrespective of the sediment. Thus, both water and sediment are sources of lead for this plant. Samples grown in uncontaminated water and sediment accumulated lead concentrations of 5.2 mg/kg while those in contaminated water or sediment accumulated up to 160.9 mg/kg.

Ruhling & Tyler (1970) analysed the lead content of mosses *Hypnum cupressiforme* and *Hylocomium splendens* in different regions of Scandinavia to monitor fall-out of industrial lead. They found significantly higher levels of lead in *H. splendens* from southern Sweden compared with northern Scandinavia (11 mg/kg) and also higher levels in south-west (90 mg/kg) than in south-east Sweden (52 mg/kg). The same pattern was found in lead levels of *H. cupressiforme* between different areas of southern Sweden. The authors eliminated all other possible sources of lead than anthropogenic ones.

Edelman et al. (1983) analysed earthworms *(Lumbricus rubellus)* and soil samples, for lead content, near a zinc-smelting complex. Levels in soil ranged from 14 to 430 mg/kg dry weight and in worms from 9 to 670 mg/kg dry weight. Although there was a significant correlation between distance from smelter and levels of lead in soil and worms, it was not as strong a relationship as for cadmium or zinc. Soil lead content, soil pH, and soil organic matter together accounted for 70% of variance in worm lead uptake. The authors found higher lead levels in worms from soil of a lower pH and lower organic matter content. Lead was estimated after clearing the worms of gut contents.

Bengtsson & Rundgren (1984) analysed ground-living invertebrates, such as spiders, harvestmen, slugs, beetles, and ants, from metal-polluted forest soils, at varying distances from a Swedish brass mill. Mean lead levels were significantly higher in most of the species within 650 m of the mill. Litter levels of lead of 600-1000 mg/kg were found, dependent on the distance from the mill. Lead levels in litter were 20-30 times less than zinc or copper levels.

Roberts et al. (1978) found significantly higher lead levels in surface soil, vegetation, and invertebrates at two abandoned non-ferrous mine spoil tips than in control areas. The two areas showed lead at 8430 and 14 010 mg/kg dry weight soil, 120 and 249 mg/kg dry weight vegetation, and 61.9 and 81.7 mg/kg dry weight invertebrates. Four species of small mammals *(Microtus agrestis, Apodemus sylvaticus,*

Clethrionomys glareolus, and *Sorex araneus)* showed significantly higher levels of lead when trapped in the contaminated areas. The highest levels in *M. agrestis* were 45.3 and 42.8 mg/kg fresh weight, for the two areas. When tissues of *A. sylvaticus* were analysed, kidney, liver, bone, and brain contained significantly higher levels than controls. Lead levels of 352 and 189 mg/kg dry weight for bone compared with 11.5 and 21.1 mg/kg for the two control areas. Muscle residues were not significantly different between areas. Similar results were found in the tissues of *A. sylvaticus* living on smelter waste (Johnson et al., 1978). Surface soil contained 4030 mg lead/kg (control 76.1 mg/kg dry weight) and bone levels were 672 mg/kg dry weight (control 34.2 mg/kg).

Cloutier et al. (1986) assayed the lead content of tissues from meadow voles *(Microtus pennsylvanicus),* living on nickel or uranium mine tailings. Soft tissue levels of lead were below detection limits in most cases. Bone levels of lead were slightly elevated at the uranium site, but not significantly. The highest levels were 21.9 mg/kg dry weight in sub-adults and 23 mg/kg in adults. No sex or age differences were reported. There was a rise in bone lead levels between winter and the following autumn in the lead-rich area (the uranium site), but a fall over the same period at the nickel site and control site.

Smith & Rongstad (1982) determined lead concentrations in the whole body of *Peromyscus maniculatus* and *M. pennsylvanicus* from an active zinc-copper mine and a proposed zinc-copper mine. *P. maniculatus* from the proposed mining site showed lead concentrations of the same order as in a non-mining control area. From the mining site, there were consistently higher concentrations for both sexes and ages, juveniles and adults. *M. pennsylvanicus* in the mining site showed no elevation in lead content over controls.

4.2.4 Lead shot

Appraisal

Lead shot taken by birds into their gizzards is a source of severe lead contamination. In organs, high levels of lead are found in blood, kidney, liver, and bone.

Mudge (1983) analysed for lead 1620 livers and 1871 wing bones from 23 species of British waterfowl (shot or found dead). The highest levels of lead in the liver were found in birds with ingested pellets in the gizzard. The species with the highest levels, excluding those birds without ingested pellets in the gizzard, were gadwall (11.3-22.0 mg/kg dry weight), mute swan (11.6-32.7 mg/kg), Bewick's swan (73.0-109.9 mg/kg), and greylag goose (57.2-61.9 mg/kg). Of the 14 species of duck analysed for lead in the wing bone (and not containing lead shot at the time of sampling), the highest levels were in mallard (<5.0-472.9 mg/kg dry weight), teal (<5.0-298.8 mg/kg), and wigeon (<5.0-175.9 mg/kg). The author also assayed 63 blood samples from four species of waterfowl; the highest mean blood lead levels were in whooper swan (4.6 mg/litre) and Bewick's swan (8.3 mg/litre).

Analysis for lead of mute swan blood samples, taken from swans from contaminated and uncontaminated areas in the United Kingdom, has revealed large differences between areas (NCC, 1981) (see also section 8.2). The highest level reported was from the River Trent, Nottingham (3.75 mg/litre), and the lowest level was 0.08 mg/litre from the Abbotsbury swannery, Dorset. Simpson et al. (1979) analysed various organs of lead-poisoned mute swans found dead. The highest lead levels were found in the kidney (350-6550 mg/kg dry weight), liver (51-206 mg/kg), and bone (212-1255 mg/kg). These levels compared with 'healthy' control swan levels of 1-77 mg/kg kidney, 1-11 mg/kg liver, and 21-41 mg/kg bone.

Anderson (1975) examined about 1500 waterfowl dying at Rice Lake, Illinois, USA. When 96 lesser scaup, of which 75% had at least one lead pellet in the gizzard, were analysed for lead, the mean levels were 46 mg/kg, 66 mg/kg, and 40 mg/kg for liver, kidney, and wing bone, respectively.

5. TOXICITY TO MICROORGANISMS

Appraisal

In general, inorganic lead compounds are of lower toxicity to microorganisms than are trialkyl- and tetraalkyllead compounds. Tetraalkyllead becomes toxic by decomposition into the ionic trialkyllead.

One of the most important factors which influence the aquatic toxicity of lead is the free ionic concentration, which affects the availability of lead for organisms. The toxicity of inorganic lead salts is strongly dependent on environmental conditions, such as water hardness, pH, and salinity, a fact which has not been adequately considered in most toxicity studies.

There is evidence that tolerant strains exist and that tolerance may develop in others.

5.1 Toxicity of Lead Salts

Bringmann & Kuhn (1959a) reported a toxic threshold for lead, as lead nitrate, of 1.3 mg/litre for the bacterium *Escherichia coli,* related to cell numbers produced. Lead, as the nitrate or bromide, had little effect on the growth of the human skin bacterium *Micrococcus luteus* at a level of 600 μg/litre over 48 h (Tornabene & Edwards, 1973). The latter authors recultured bacteria after the lead treatment, with inocula transferred to fresh medium after 48 h of growth. After 20 days of continuous growth, the cellular yield had decreased to less than half that of a control culture. The pigmentation of this characteristically yellow bacterium was reduced by this time. Electron-microscopic examination of these cells indicated that cytoplasmic material was leaking out. Lead is largely concentrated in the cell membranes of bacteria, and could be seen as electron-dense inclusions; membrane breakdown was usually seen in the area of lead inclusions.

Gray & Ventilla (1971) cultured the ciliate *Cristigera* sp. on a diet of bacteria *(Pseudomonas* sp.), both organisms having been isolated from beach sand. Lead nitrate added to the cultures reduced the growth rate, but did not inhibit growth at between 0.1 and 0.3 mg/litre. The result was significant at the 5% level.

Monahan (1976) reported a 50% reduction in cell numbers of the freshwater alga *Selenastrum capricornutum* after 7 days exposure to lead in the culture medium at a concentration of 0.5 mg/litre medium. Increasing the pH of the medium from acidic to alkaline levels reduced the toxicity of lead to the alga. Christensen et al. (1979) used the same freshwater alga and a second alga, *Chlorella stigmatophora,* cultured in artificial sea water, in a study of the effects of inorganic lead, alone and in combination with other metals. *Selenastrum* was cultured in standard algal assay medium (SAAM). In an initial range-finding test, *Selenastrum* and *Chlorella* were cultured with lead, as lead acetate, in solution at concentrations of 0.01, 0.1, 1.0, 10.0, 100.0, and 1000.0 mg/litre. The effects were assessed in terms of total cell volume, relative to controls, after 13 days for *Selenastrum* and 24 days for *Chlorella*. *Selenastrum* was slightly stimulated by lead acetate at 0.01 mg/litre, with a relative cell volume of 1.05. At 0.1 mg lead/litre, *Selenastrum* showed a reduced cell volume of 0.87 relative to controls, and, at 1.0 mg lead/litre, a cell volume ratio of 0.12. At concentrations of 10.0 mg/litre or more, lead killed the algal cells. Only at the highest exposure of 1000 mg/litre was there any visible precipitation of the lead acetate. *Chlorella* was also killed by lead acetate at concentrations of 10.0 mg/litre or more in the artificial sea water. At 1.0 mg/litre, lead reduced the cell volume of *Chlorella* to 0.71 relative to controls. At lower lead concentrations, there was a stimulation of the alga with cell volumes of 2.09 and 1.60 relative to controls for exposures to 0.01 and 0.1 mg lead/litre, respectively. In both *Selenastrum* and *Chlorella*, lead increased the average cell volume of the algal cells significantly at the same time as it reduced the growth rate. In these experiments, *Selenastrum* was exposed to lead concentrations varying between 0.09 and 1.44 mg/litre. Over this range, growth rate decreased from 1.12 mm^3/litre per day at 0.09 mg lead acetate/litre to 0.5 mm^3/litre per day at 1.44 mg lead acetate/litre. The average volume of individual cells increased from 62 μm^3 to 91 μm^3 over the same dose range of lead in the culture medium. When *Chlorella* was exposed to a range of lead concentrations in artificial sea water between 0.36 and 5.76 mg/litre, the growth rate declined from 0.68 mm^3/litre per day to 0.4 mm^3/litre per day, and the average cell volume increased from 24 μm^3 to 61 μm^3. Values for EC_{50} for cell volume were calculated at 140 μg lead acetate/litre for *Selenastrum* and 700 μg/litre for *Chlorella*. The authors suggest that the discrepancy between their own result for *Selenastrum* and that reported by Monahan (1976) might be due to the greater concentration (by about five times) of dissolved salts in his culture medium.

Culturing the algae in a medium containing combinations of metals showed that the presence of manganese or copper reduces the toxicity of lead to these organisms (Christensen et al., 1979). Prasad & Prasad (1982) exposed three freshwater green algae (*Ankistrodesmus falcatus, Scenedesmus obliquus,* and *Chlorococcum* sp.) to lead chloride at concentrations of 0 to 10 mg lead/litre, and measured growth on the 10th day after inoculation using an optical density method. There was no effect on growth from 0.1 to 1.5 mg lead/litre, but at 2.0 mg/litre or more, there was inhibition of growth in all three species. At 10 mg lead/litre, *A. falcatus* was killed, and *S. obliquus* and *Chlorococcum* sp. were reduced to 9% and 15%, respectively, of the mass of controls.

Hongve et al. (1980) exposed a natural phytoplankton community to concentrations of an unspecified inorganic lead salt ranging from 5×10^{-7} to 5×10^{-4} mol/litre. The community of organisms, isolated from lake water, consisted mainly of diatoms: *Tabellaria flocculosa* (53% by volume), *Synedra* sp. (13%), and *Asterionella formosa* (7%). Other important constituent species were *Cryptomonas* spp., *Rhodomonas minuta* variety *lacustris, Dinobryon divergens,* small species of chryptomonades, *Gymnodinium* sp., and *Mallomonas* sp. The authors monitored photosynthetic activity as uptake of ^{14}C-labelled hydrogen carbonate over a 20-h incubation. Photosynthetic carbon fixation was reduced in a dose-dependent manner throughout the exposure range of lead in the medium; the reduction was 90% relative to control cultures at the highest exposure of 5×10^{-4} mol lead/litre. The addition of lake sediment to treated cultures reduced the toxic effect of lead; addition to control cultures increased the photosynthetic carbon uptake by 19%. A similar reduction in the toxic effects of lead was seen on adding organic matter filtered out of the lake water and after adding the chelating agent nitrilotriacetic acid (NTA) at non-toxic levels. Neither of these two variables affected photosynthesis in control cultures. The NTA had the greatest effect on lead toxicity, virtually eliminating the effect of lead on photosynthesis.

Persoone & Uyttersprot (1975) examined the effects of lead chloride on reproduction in the marine ciliate *Euplotes vannus* by estimating the number of generations produced after culture for 48 h. The *Euplotes* cultures were exposed to 0.001, 0.01, 0.1, 1, 10, or 100 mg lead chloride/litre. Reproduction was unaffected by lead at concentrations up to 0.1 mg/litre in the culture medium. At 1 mg/litre, the lead caused approximately 15% inhibition of reproduction and, at 10 mg/litre, 30% inhibition. At 100 mg lead chloride/litre, all the ciliates died.

Hessler (1974) exposed a marine unicellular green flagellate alga *(Platymonas subcordiformis)* to lead chloride at concentrations of 100, 500, and 1000 mg lead/litre of sea-water medium. There was precipitation of lead from solution and these doses gave corresponding values for lead in solution of 2.5, 10, and 60 mg/litre, respectively. Log-phase cells, growing exponentially, were more sensitive to lead than stationary-phase cells. At 2.5 and 10 mg/litre, lead retarded population growth by delaying cell division and daughter cell separation. A concentration of 60 mg/litre caused complete inhibition of growth and cell death. Normal wild-type cells were more sensitive to lead than either cells sheared of their flagellae or cells of a mutant without flagellae.

Hessler (1975) exposed *Platymonas* to the same range of lead concentrations but in the presence of mutagenic agents (ultraviolet irradiation or nitroguanidine). High levels of mutation were found but were not increased in the presence of lead.

Malanchuk & Gruendling (1973) estimated EC_{50}s for reduction in $^{14}CO_2$-fixation in freshwater algae after exposure to lead nitrate. Results were extrapolated from a graph of milligrams lead per litre plotted against radioactivity per milligram dry cell weight. Results varied with the size of the inoculum (cell density). For the cyanophyte (blue-green alga) *Anabaena* sp. the EC_{50} was 15 mg/litre

(1- and 2-ml samples) or 26 mg/litre (4-ml sample). For the chlorophytes *Chlamydomonas reinhardti* and *Cosmarium botrytis,* a desmid, EC_{50}s were 17 and 5 mg/litre, respectively. The chrysophyte *Navicula pelliculosa* showed EC_{50}s of 17 mg/litre (1- and 2-ml samples) and 28 mg/litre (4-ml sample). However, another chrysophyte, *Ochromonas malhamensis,* was not inhibited by up to 30 mg lead/litre.

Whitton (1970) investigated the effects of lead chloride on a variety of species of filamentous green alga isolated from flowing streams in northern England; some were from metal-polluted streams and some from unpolluted ones. Results were expressed semi-quantitatively and a tolerance index was determined in terms of lead concentration. This index is a geometric mean of codings indicating minimal and maximal effects. These values varied between 3 and 60 mg lead/litre. The most sensitive species, *Cladophora,* and the least sensitive species, *Microspora,* were unusual; all others tested gave values between 17 and 46 mg/litre.

Bringmann & Kuhn (1959b) reported a toxic threshold of 2.5 mg/litre for the green alga *Scenedesmus* (related to cell division) and of 1.25 mg/litre for the protozoan *Microregma* (related to feeding).

Babich & Stotzky (1983) observed that hard water protected *Tetrahymena pyriformis* from the effects of lead salts. Gray & Ventilla (1973) exposed a sediment-living, bacterivorous ciliate protozoan, *Cristigera* sp., to concentrations of 0.15 or 0.3 mg lead nitrate/litre for 4 to 5 h. Lead reduced growth rates by 8.5% and 11.8% for the two doses, respectively. Analysis of variance showed the effect to be significant at the 1% level. Apostol (1973) used another ciliate protozoan, *Paramecium caudatum,* in acute and long-term tests to examine the toxicity of lead acetate. In the 5-h acute test, *Paramecium* showed a sharp threshold of toxic response at around 1000 mg lead acetate/litre, with survival times dropping steeply from >300 min to 5 min or less. In chronic tests over 14 days, growth of the population was delayed by lead acetate at 1, 10, and 100 mg/litre. Peak population numbers were progressively reduced as lead concentrations increased. At the beginning of the test, the median survival span at 1000 mg/litre was <5 min, whereas at the end of the test, the survival span was >5 h at the same concentration. It is clear, therefore, that there is considerable individual variation in the population and scope for adaptation in the wild.

Ruthven & Cairns (1973) determined the minimal lethal concentration and maximum tolerated concentration of lead, as lead nitrate, for six different species of freshwater algae and protozoans. Two species, *Peranema* and *Euglena gracilis,* tolerated 1000 mg lead/litre (nominal concentration). *Blepharisma* tolerated 42 mg/litre; *Tetrahymena* and *Paramecium multimicronucleatum* tolerated 24 mg/litre, while *Chilomonas* tolerated only 5.6 mg/litre. Minimal lethal concentrations for the four more sensitive species ranged from 56 mg/litre to >100 mg/litre.

Rosenweig & Pramer (1980) examined the effects of lead nitrate on seven species of nematode-trapping fungi from soil. Mycelial growth was reduced in two species at lead concentrations of 100 mg/litre, and in all but one species at 300 mg/litre. Reduced capacity to produce traps (rings of mycelium which capture nematode worms) was correlated with reduced growth, except in the case of one species where growth was inhibited with no effect on trap production. Increasing pH reduced the toxicity of lead to the fungi *Aspergillus niger* (Babich & Stotzky, 1979), *Achyla* sp. and *Saprolegnia* sp. (Babich & Stotzky, 1983). Babich & Stotzky (1979, 1983) noted that the presence of carbonate or phosphate ions reduced the toxic effect of lead on *Aspergillus* and *Fusarium* growth, presumably by precipitating lead from the medium.

Crist et al. (1985) collected and dried green leaves from a variety of tree species representative of central hardwood forests of the USA. The leaf mixtures were treated with lead sulfate to give concentrations of lead in the leaf litter ranging from 0 to 1000 mg/kg and incubated in laboratory microcosms. Replicates were treated with different amounts of sulfuric acid to give pH values in the incubates of between 3 and 5. Lead, at these concentrations, had no effect on leaf decomposition at any of the pH values tested.

5.2 Toxicity of Organic Lead

Roderer (1980) investigated the effects of tetraethyllead on a flagellated alga, *Poterioochromonas malhamensis*. After 3 days of culture in darkness, there were no toxic effects of tetraethyllead even at concentrations of 0.3 mmol/litre. At 0.25 mmol/litre in light, all cells were killed by tetraethyllead. At concentrations below 0.25 mmol/litre, there was a dose-related effect on growth, mitosis, and cytokinesis, resulting in the formation of giant polyploid cells. Tetraethyllead was converted to a highly toxic derivative in light with, or without, cells present. This toxic compound was produced in toxic amounts within 3 to 6 h of illumination, but reached a maximum concentration after 24 to 32 h. Free radicals, which are produced during the photolysis of tetraethyllead, were shown not to be responsible for the toxicity. Tetraethyllead is removed from water by aeration because of its low water solubility and high volatility. In the presence of light, this process is counteracted by the formation of a stable, water-soluble material, which is toxic to algae. The authors identified the toxic product as triethyllead.

Marchetti (1978) investigated the effects of tetraalkyllead in natural sea water on mixed coastal marine bacteria using the biological oxygen depletion method in a respirometer. Two commercial products were used: tetramethyllead TML-CB and tetraethyllead TEL-CB. Tetraalkyllead solutions were prepared by adding 2 ml of each product to 1 litre of filtered natural sea water and slowly stirring magnetically for 1 h. The upper quarter of the solution was used for preparing experimental dilutions. The lag phase was related to the TML-CB lead concentration up to lead concentrations in water of 3.2 mg/litre and to the TEL-CB lead concentration up to 0.16 mg/litre. There was also a relationship between lead concentration and respiration rate, starting from 0.36 and 0.08 mg/litre, respectively, for TML-CB and TEL-CB. Below these concentrations there was no significant effect on either lag phase or oxygen consumption. The EC_0, EC_{50}, and EC_{100} over 48 h were 0.9, 1.9, and 4.5 mg/litre, respectively, for TML-CB and 0.08, 0.2, and 2.0 mg/litre for TEL-CB. TML-CB is less toxic to bacteria than TEL-CB, on the basis of total lead content, even if the presence of some toluene in the TML formulation is taken into account. The author speculates that the difference in toxicity depends on the different speed of transformation from the tetraalkyl form, scarcely soluble and toxic, to the more soluble and less toxic trialkyl form. Tests with the same preparation on the photosynthesis of the alga *Dunialiella tertiolecta* gave an EC_0, an EC_{50}, and an EC_{100} of 0.45, 1.65, and 4.5 mg/litre, respectively, for TML-CB and 0.1, 0.15, and 0.3 mg/litre for TEL-CB.

Silverberg et al. (1977) exposed the freshwater algae *Scenedesmus quadricaudata, Ankistrodesmus falcatus,* and *Chlorella pyrenoidosa* to tetramethyllead. Tetramethyllead is not soluble in water and is volatile. The compound was biologically generated in a reaction vessel using trimethyllead acetate and *Aeromonas* sp. or indigenous microorganisms in Hamilton Harbour water and sediment. When tetramethyllead was detected in air drawn off the reaction vessel, this was bubbled through cultures of the algae. Exposure was, because of the nature of the material, momentary because of conversion to trimethyllead. The primary productivity of the cultures was estimated using ^{14}C-hydrogen carbonate uptake, and growth was estimated using both cell dry weight and counts of cell numbers. Cells were also harvested and fixed for electron-microscopic evaluation. Although the exposure cannot be estimated exactly, the authors estimate that <0.5 mg of tetramethyllead was passed through the cultures during the course of the 7-day study. *Chlorella* was the most sensitive of the three organisms showing a decrease of 74% in growth and 83% in photosynthesis. *Scenedesmus* showed a 32% decrease in growth and 85% decrease in photosynthesis; the corresponding figures for *Ankistrodesmus* were 32% and 49%, respectively. The cultures showed loss of green coloration, the green becoming semitransparent yellow with time. Cells were enlarged and clumped into masses. After electron-microscopic examination, it could be seen that the chloroplasts were most affected by the tetramethyllead. Lead was detected inside the cells using electron-microscopic analysis. The authors state that tetramethyllead is twice as toxic as trimethyllead acetate and 20 times more toxic than lead nitrate for the same organisms.

Roderer (1983) found that compounds used to alleviate lead poisoning in man (Na_2EDTA, EDTA, DPA, DIZO, BAL) increased, rather than decreased, the effects of inorganic and triethyllead on the unicellular alga *Poterioochromonas malhamensis.* In a later, more comprehensive investigation of factors affecting the toxicity of triethyllead to *Poterioochromonas* (Roderer, 1986), the author studied the protective action of thiol compounds, vitamins, trace elements, and other agents. None of the tested thiol or disulfide compounds protected the alga from triethyllead. Two vitamins, tocopheryl acetate and ascorbic acid, one trace element, zinc, and ATP, cyclic AMP, and concanavalin A, together with some combinations of agents, markedly suppressed the growth-inhibiting effects of triethyllead. Zinc was the most effective single agent, increasing growth of the algal cultures by 70 times in the presence of triethyllead at 10^{-5} mol/litre. A combination of 10 essential trace elements was even more effective and almost totally eliminated the toxic effect of the lead compound.

6. TOXICITY TO AQUATIC ORGANISMS

Lead is unlikely to affect aquatic plants at levels that might be found in the general environment.

In the form of simple salts, lead is acutely toxic to aquatic invertebrates at concentrations between 0.1 and >40 mg/litre for freshwater organisms and between 2.5 and >500 mg/litre for marine organisms. The 96-h LC_{50} for fish varies between 1 and 27 mg/litre, in soft water, and between 440 and 540 mg/litre, in hard water, for the same species. The higher values for hard water represent nominal concentrations. Available lead measurements suggest that little of the total lead is in solution in hard water. Lead salts are poorly soluble in water, and the presence of other salts reduces the availability of lead to organisms because of precipitation. Results of toxicity tests should be treated with caution unless dissolved lead is measured.

There is little information on the effects of organic lead complexes. Sublethal effects have been reported.

6.1 Toxicity to Aquatic Plants

Appraisal

There is little evidence for effects of lead on aquatic plants at concentrations below 1 to 15 mg/litre. Many studies of aquatic plants have been made in sediment-free systems. However, the addition of uncontaminated sediment reduces the toxicity of lead to aquatic plants by reducing its availability.

Van der Werff & Pruyt (1982) exposed four aquatic plants, *Elodea nuttallii, Callitriche platycarpa, Spirodela polyrhiza,* and *Lemna gibba,* to concentrations of lead nitrate of up to 10^{-5} mol lead/litre for 70 to 73 days. There was no observable toxicity, and growth rates were unaffected. Brown & Rattigan (1979) exposed the aquatic macrophyte *Elodea canadensis* (Canadian pond-weed) and the free-floating duckweed *Lemna minor* to a range of lead acetate concentrations for 28 days and 14 days, respectively. The authors assessed damage to the plants visually on a coded scale from 0 (no damage) to 10 (complete plant kill). They reported that concentrations of 136 and 16.3 mg/litre produced 50% damage to the two plant species, respectively. In a separate experiment, they exposed *Elodea* to lead for 24 h in the dark, and then measured oxygen evolution in the light. Levels of 47.6 and 99 mg lead/litre reduced photosynthetic oxygen evolution by 50% and 90%, respectively. Kay et al. (1984) exposed the water hyacinth *Eichhornia crassipes* to lead nitrate concentrations of 0.5 to 5.0 mg/litre for 6 weeks. There was no observed effect on root development, leaf colour, development of new plantlets, flowering, or total plant growth.

Stanley (1974) determined EC_{50}s for various growth parameters of Eurasian watermilfoil *(Myriophyllum spicatum)* exposed to lead (salt unspecified). Plants were grown in soil with water above. The EC_{50} for root weight was 363 mg/litre, for shoot weight was 808 mg/litre, for root length was 767 mg/litre, and for shoot length was 725 mg/litre. The effects of adding the lead to the soil as opposed to the water were investigated. There was less effect with lead added to the soil because of adsorption to soil particles. There was a ratio of 1.43 between root growth when lead was added to soil over that when lead was added to water, following exposure to 20.7 mg lead/litre. For exposure to 207 mg lead/litre, the corresponding ratio was 1.88.

6.2 Toxicity to Aquatic Invertebrates

Appraisal

The results of experiments on the toxicity of lead salts to aquatic invertebrates are difficult to interpret due to the variations in experimental conditions and the lack of a standardized method for determining lead concentrations in water. In most studies, concentrations of lead in water are nominal; the contribution to toxicity of factors, such as pH, water hardness, anions, and complexing agents cannot be fully evaluated.

In communities, some populations of organisms are more sensitive than others, and community structure may be adversely affected by lead contamination. However, populations from polluted areas can show more tolerance to lead than those from non-polluted areas. In other organisms, adaptation to hypoxic conditions can be hindered by high lead concentrations.

There is information on the toxicity of lead salts to aquatic invertebrates, but little information on the effects of organic lead compounds. The toxicity of lead to aquatic invertebrates is summarized in Tables 3 and 4.

6.2.1 Toxicity of lead salts

Cleland (1953) found that lead nitrate at a concentration of 6×10^{-4} mol lead/litre suppressed the development of a fertilization membrane elevation in eggs of the sea urchin *(Psammechinus miliaris)*. Subsequent cleavage of the fertilized egg was generally normal.

Watling (1983b) reported that the larvae of the oyster *Crassostrea gigas* grew less well, over a 14-day exposure period, with lead nitrate in the water at 0.01 or 0.02 mg/litre. The exposed larvae showed a mean length of 5.0 and 5.3 mm, respectively, in solutions of 0.01 and 0.02 mg lead/litre, after 14 days, compared to 6.3 mm for the controls. After a further 14 days in clean water, most of the reduction in size had been recovered. The treated larvae were 8.0 and 7.8 mm mean length

Table 3. Toxicity of lead salts to aquatic invertebrates

Organism	Life-stage	Flow/stat[a]	Temp. (°C)	Alkalinity[c]	Hardness[c]	pH	Salt	Parameter	Water concentration (mg/litre)	Reference
American oyster (*Crassostrea virginica*)		stat	25-27	25[d]			nitrate	48-h LC_{50}	2.45 (2.2-3.6)	Calabrese et al. (1973)
		stat	25-27	25[d]			nitrate	48-h LC_0	0.5	
		stat	25-27	25[d]			nitrate	48-h LC_{100}	> 6.0	
Hard clam (*Mercenaria mercenaria*)		stat	25-27	25[d]			nitrate	48-h LC_{50}	0.78 (0.72-0.80)	Calabrese & Nelson (1974)
		stat	25-27	25[d]			nitrate	48-h LC_{100}	1.20	
Softshell clam (*Mya arenaria*)		stat	21.5-22.5	29-31[d]		7.8-8	nitrate	48-h LC_{50}	> 50	Eisler (1977)
		stat	21.5-22.5	29-31[d]		7.8-8	nitrate	96-h LC_{50}	27	
		stat	21.5-22.5	29-31[d]		7.8-8	nitrate	168-h LC_{50}	8.8	
Cockle (*Cardium edule*)	adult	stat	15				nitrate	48-h LC_{50}	> 500	Portmann & Wilson (1971)
Pink shrimp (*Pandalus montagui*)	adult	stat	15				nitrate	48-h LC_{50}	375	Portmann & Wilson (1971)
Neanthes arenacoe-dentata	juvenile	stat				7.8	acetate	96-h LC_{50}	> 7.5[e]	Reish et al. (1976)
	adult	stat				7.8	acetate	96-h LC_{50}	> 10[e]	
	juvenile	stat				7.8	acetate	28-day LC_{50}	2.5[e]	Reish et al. (1976)
	adult	stat				7.8	acetate	28-day LC_{50}	3.2[e]	
Capitella capitella	larva	stat				7.8	acetate	96-h LC_{50}	1.2[e]	Reish et al. (1976)
	adult					7.8	acetate	96-h LC_{50}	6.8[e]	
	adult					7.8	acetate	28-day LC_{50}	1.0[e]	Reish et al. (1976)
Crab (*Scylla serrata*)		stat	26.5-29.5			7.0-7.2	nitrate	96-h LC_{50}	> 370	Krishnaja et al. (1987)

Table 3 (contd).

Mussel (*Lamellidens marginalis*)	stat	28-32	7.7-11.7	32-38	7.0-7.3	nitrate	48-h LC_{50}	> 40	Subbaiah et al. (1983)
Freshwater crab (*Oziotelphusa senex senex*)	stat	28-32	7.7-11.7	32-38	7.0-7.3	nitrate	48-h LC_{50}	> 40	Subbaiah et al. (1983)
Snail (*Pila globosa*)	stat	28-32	7.7-11.7	32-38	7.0-7.3	nitrate	48-h LC_{50}	> 40	Subbaiah et al. (1983)
Copepod (*Cyclops abyssorum*)	stat	9.5-10.5	0.58 meq/litre		7.2	acetate	48-h LC_{50}	5.5 (4.0-7.7)	Baudouin & Scoppa (1974)
Copepod (*Eudiaptomus padanus*)	stat	9.5-10.5	0.58 meq/litre		7.2	acetate	48-h LC_{50}	4.0 (2.5-6.4)	Baudouin & Scoppa (1974)
Water flea (*Daphnia magna*)	stat		41-50	44-53	7.4-8.2	chloride	48-h LC_{50}	0.45^{f}	Biesinger & Christensen (1972)
	stat		41-50	44-53	7.4-8.2	chloride	21-day LC_{50}	0.3 (0.236-0.381)	
	stat	11.5-14.5	390-415	235-260	7.4-7.8	acetate	24-h LC_{50}	4.89 (4.19-5.89)	Khangarot & Ray (1987)
	stat	11.5-14.5	390-415	235-260	7.4-7.8	acetate	48-h LC_{50}	3.61 (2.83-4.4)	Khangarot & Ray (1987)
Water flea (*Daphnia hyalina*)	stat	9.5-10.5	0.58 meq/litre		7.2	acetate	48-h LC_{50}	0.60 (0.41-0.89)	Baudouin & Scoppa (1974)
Amphipod (*Gammarus pseudolimnaeus*)	flow	15	40-44	44-48	7.1-7.7	nitrate	96-h LC_{50}	0.124	Spehar et al. (1978)

Table 3 (contd).

Organism	Life-stage	Flow/stat[a]	Temp. (°C)	Alkalinity[c]	Hardness[c]	pH	Salt	Parameter	Water concentration (mg/litre)	Reference
Crayfish (*Austropotamobius pallipes pallipes*)		flow[b]	15-17			7.0	chloride	96-h LC_{50}	2.6	Boutet & Chaisemartin (1973)
		flow[b]	15-17			7.0	chloride	30-day LC_{50}	1.5	
		flow[b]	15-17			7.0	chloride	30-day LC_{50}	0.9[e]	
Crayfish (*Orconectes limosus*)		flow[b]	15-17			7.0	chloride	96-h LC_{50}	3.3	Boutet & Chaisemartin (1973)
		flow[b]	15-17			7.0	chloride	30-day LC_{50}	1.7	
		flow[b]	15-17			7.0	chloride	30-day LC_{50}	0.9[e]	
Midge (*Tanytarsus dissimilis*)	egg/larva	stat	21-23	43.9	46.8	7.5	nitrate	10-day LC_{50}	0.258	Anderson et al. (1980)
Mayfly (*Epherella grandis*)	larva	flow				7.0-7.2	nitrate	14-day LC_{50}	3.5	Nehring (1976)
Stonefly (*Pteronarcys californica*)	larva	flow				7.0-7.2	nitrate	14-day LC_{50}	> 19.2	Nehring (1976)

[a] Stat = static conditions (water unchanged for duration of test); flow = flow-through conditions (lead concentration in water continuously maintained).
[b] Intermittent flow-through conditions.
[c] Alkalinity and hardness expressed as mg/litre $CaCO_3$.
[d] These figures are values for salinity (expressed as $^{o}/_{oo}$), not alkalinity.
[e] With a food source.
[f] Water fleas were fed during test.

Table 4. Toxicity of organolead to aquatic invertebrates

Organism	Mean length (mm)	Mean weight (g)	Flow/stat[a]	Temp. (°C)	Salinity (°/oo)	Compound[b]	Parameter	Water concentration (mg/litre)	Reference
Mussel	64	28.5	flow	15	34.9	TML	96-h LC_{50}	0.27	Maddock & Taylor (1980)
(*Mytilus edulis*)	64	28.5	flow	15	34.9	TEL	96-h LC_{50}	0.1	Maddock & Taylor (1980)
	64	28.5	stat	15	34.9	TriML	96-h LC_{50}	0.5	Maddock & Taylor (1980)
	64	28.5	stat	15	34.9	TriEL	96-h LC_{50}	1.1	Maddock & Taylor (1980)
Brown shrimp	48	1.1	flow	15	34.9	TML	96-h LC_{50}	0.11	Maddock & Taylor (1980)
(*Crangon crangon*)	48	1.1	flow	15	34.9	TEL	96-h LC_{50}	0.02	Maddock & Taylor (1980)
	48	1.1	stat	15	34.9	TriML	96-h LC_{50}	8.8	Maddock & Taylor (1980)
	48	1.1	stat	15	34.9	TriEL	96-h LC_{50}	5.8	Maddock & Taylor (1980)

[a] Stat = static conditions (water unchanged for duration of test); flow = flow-through conditions (lead concentration in water continuously maintained).

[b] TML = tetramethyl lead; TEL = tetraethyl lead; TriML = trimethyl lead chloride; TriEL = triethyl lead chloride.

for the two dose levels, and the controls were 8.2 mm long. The author also reported that lead, at both 0.01 and 0.02 mg/litre, reduced the numbers of larvae settling and delayed the peak settlement time of the population.

Calabrese et al. (1973) found that the EC_{50} of lead chloride for the development of larvae of the American oyster was 2.45 mg/litre; the EC_0 was 0.5 mg lead/litre. Lead nitrate is more toxic to the hard clam (*Mercenaria mercenaria*) than to the American oyster (*Crassostrea virginica*) (Calabrese & Nelson, 1974) (Table 3). Coombs (1977) exposed batches of 20 mature mussels, *Mytilus edulis*, of shell length 6-7 cm, to lead (added to the water as nitrate or complexed with citrate, humic and alginic acids, or pectin). The author showed that the uptake of lead was increased by complexation, citrate being the most effective complexing agent for stimulating absorption of the metal. Electron-microscopic examination of tissues showed that the mussels were able to tolerate large amounts of lead in their tissues, and to reduce its toxicity by enclosing the metal in membrane-bound vesicles. Stromgren (1982) reported that lead citrate, at water concentrations of up to 0.2 mg/litre, had no effect on the growth rate of *Mytilus edulis*.

Lead nitrate in water, at concentrations of up to 0.565 mg/litre, had no effect on the survival of the freshwater snail *Physa integra* (Spehar et al., 1978). Borgmann et al. (1978) exposed the freshwater snail *Lymnaea palustris* to various lead nitrate concentrations, in a flow-through study, ranging from 3.8 to 54 μg/litre over 120 days. There was no effect on survival at concentrations of 3.8 and 12 μg/litre, but mortality occurred at concentrations of 19 μg/litre or more. The growth rate of the survivors was not affected at lead concentrations of 19 μg/litre. The authors observed a 50% reduction in snail biomass production after exposure to lead at 36 μg/litre from hatching during the period of maximal growth.

Baudouin & Scoppa (1974) reported LC_{50} results for two species of freshwater copepod and for a water flea (Table 3). They failed to find any indication of a lethal threshold for lead. Roberts & Maguire (1976) added inorganic lead (salt unspecified) to sand, collected from the surface and sub-surface below mid-tidal level, at concentrations of 0.001, 0.1, or 1 mg/litre in sea water. The populations of various meiofauna were estimated with time, up to 410 h after adding the lead. The most affected organisms in the surface sand were harpacticoid copepods, whose numbers declined with time and increasing lead concentrations. Measurement of lead in the interstitial water of the test samples showed that much of the metal was strongly adsorbed to sand particles very early in the experiment. Only for the first 10 h, at the highest exposure, were significant amounts of lead detectable in the water (~0.1 mg/litre). Nematodes were the most sensitive organisms in sub-surface sand.

Fraser et al. (1978) collected samples of the freshwater crustacean *Asellus aquaticus* from various polluted and unpolluted sites in the basin of the River Trent, United Kingdom. The different populations were exposed for 24 h in the laboratory to lead nitrate solutions at pH 4.5 and lead concentrations of 0, 100, 250, 500, 750, 1000, and 1500 mg/litre. The authors found a log-linear relationship between lead concentration and survival of the *Asellus*. Animals less than 4 mm in length survived less well than larger animals at the higher lead concentrations. Approximately 50% of both small and large *Asellus* survived for 24 h after exposure to lead nitrate at 100 mg/litre. Those animals collected from an area with higher lead levels were more tolerant to the metal in laboratory experiments, suggesting some selection in the wild. Exposure in the wild, during 3 years of analysis, varied between 0 and 0.24 mg/litre in the high-lead area and between 0 and 0.08 mg/litre in the low-lead area.

Spehar et al. (1978) exposed the freshwater amphipod *Gammarus pseudolimnaeus* to lead nitrate solutions in lake water for 28 days. The lead caused more than 50% mortality at water concentrations of 0.136 mg/litre or more, over 4 days. By the end of the study, mortality was 60% at the lowest concentration of lead nitrate tested, 0.032 mg/litre. Survival curves showed a marked increase in slope between test concentrations of 0.067 and 0.136 mg/litre. At higher concentrations of lead nitrate, virtually all the final mortality occurred within the first 7 days of exposure.

Freedman et al. (1980) investigated the effect of lead speciation on the toxicity of the metal to the shrimp *Hyallela azteca* in artificial test media. Lead was added to the medium in association with four different molarities of phosphate, 10^{-3}, 10^{-4}, 10^{-5}, and 10^{-6} mol/litre, and at two different pH values, 6 and 8. Theoretical calculations were made of the concentration of free lead in the solutions. At pH 6, very little free lead exists at high phosphate concentrations, irrespective of the total lead concentration. Similarly, little free lead is predicted at any phosphate concentration at pH 8. Mortality figures related well to the predicted values for free lead in the various solutions. At pH 6, a total lead concentration of 5 mg/litre, and a phosphate concentration of 10^{-6} mol/litre, there was 100% mortality after 48 h. For phosphate molarities of 10^{-5} and 10^{-4} mol/litre, toxicity was progressively reduced. At the highest phosphate concentration, mortality only reached 25% after 120 h. Free lead values predicted for the same three phosphate concentrations were 2.76, 2.24, and 0.11 mg/litre, respectively. Chinnayya (1971) found that lead nitrate at 10^{-3} mol/litre in fresh water reduced the oxygen consumption of the shrimp *Caridina rajadhari* from a control level of 0.49 ml/h per g wet weight of shrimps to 0.38 ml/h per g. This concentration of lead caused no mortality over 10 days. The lowest concentration of lead nitrate causing mortality in this species was 5×10^{-3} mol/litre.

Anderson (1978) maintained the crayfish *Orconectes virilis* in natural river water, with lead acetate added to concentrations of 0.5, 1.0, or 2.0 mg lead/litre. The water was changed at 5-day intervals to maintain the lead concentration, and at 10-day intervals, the oxygen consumption of the crayfish was measured. There was a dose-related reduction in oxygen consumption after 10 days of exposure to lead acetate. After 20, 30, and 40 days of exposure, there was no difference in oxygen consumption between control and treated crayfish; the animals had acclimatized to the lead. The crayfish were found to be compensating for the effect of the lead, which reduced the capacity for oxygen uptake through the gills, by increasing the flow of water over the gill surfaces. There was a dose-related relationship between ventilation volume and lead concentration in the water over the range 0-2.0 mg/litre; the ventilation volume at 2.0 mg lead acetate/litre was 19 ml/min, compared with 12 ml/min for controls. Since the water in the test tanks was kept saturated with oxygen, the crayfish were able to restore fully their oxygen uptake.

Brown & Ahsanullah (1971) studied the effects of lead nitrate on mortality and growth of the worm *Ophryotrocha labronica* and the brine shrimp *Artemia salina.* When they were exposed to lead at 1 mg/litre, the LT_{50} was >600 h for *Ophryotrocha* and 576 h for *Artemia.* There was no significant suppression of growth rate (measured as increase in length) after exposure of the worm to 10 mg/litre for 8 days or 1 mg/litre for 10 days. However, a significant suppression of the growth of 48-h brine shrimp larvae was reported after exposure to lead nitrate at 5 and 10 mg/litre for 6 days.

Fischer et al. (1980) investigated the effects of lead chloride on the tubifex worm *Tubifex tubifex* under aerobic and hypoxic conditions. When worms were exposed for 6 days to lead chloride, at a concentration of 10 mg/litre of tap water, there was no mortality. The authors sectioned segments of the worms and measured the nuclear volume of the chloragocytes. These cells are responsible for the synthesis of haemoglobin and respond to hypoxic conditions by increasing their activity. Nuclear volume correlates with available oxygen. In aerated water, lead chloride caused an increase in nuclear volume of the chloragocytes from 68.8 μm^3, the control size, to 93.1 μm^3. Under hypoxic conditions, control nuclear volume increased to 137.9 μm^3, but in lead-treated animals increased only to 99.6 μm^3. This physiological response in compensating for hypoxia is essential to this animal in its normal environment, where large changes in available oxygen will be commonplace.

Biesinger & Christensen (1972) found that reproductive impairment was a more sensitive measure of the toxicity of lead chloride to water fleas (*Daphnia magna*) than survival. They determined an EC_{16} and EC_{50} of 30 and 100 μg lead/litre, respectively, for a 3-week exposure.

Warnick & Bell (1969) exposed nymphs of stonefly *(Acroneuria lycorias)*, mayfly (*Ephemerella subvaria*), and caddisfly (*Hydropsyche betteni*) to lead sulfate in static bioassays. They reported 50% survival times of >14 days at 64 mg/litre, 7 days at 16 mg/litre, and 7 days at 32 mg/litre. There was a considerable decrease in the metal concentrations in solution over the 2-week experimental period, and the authors considered that nominal concentrations were unreliable after 96 h. Spehar et al. (1978) found no effect of their highest dose of 0.565 mg lead nitrate/litre on the survival of nymphs of stoneflies or caddisflies (*Pteronarcys dorsata, Hydropsyche betteni, Brachycentrus* sp., and *Phemerella* sp.).

Anderson et al. (1980) exposed the chironomid midge, *Tanytarsus dissimilis*, to lead nitrate during two different stages of its life-cycle. Exposure started with the eggs and continued for 10 days, during which time the larvae had emerged. The average LC_{50} from two tests was 0.258 mg/litre. No significant effect on the growth of surviving larvae was found until the LC_{50} concentration was exceeded. The authors emphasized that this species is particularly sensitive to heavy metals. Chironomid midges are extremely plentiful in lakes and streams, and represent a major food source for fish.

6.2.2 Toxicity of organic lead

Marchetti (1978) determined, in 48-h tests, no-observed-effect levels (micrograms per litre) for tetraethyl- and tetramethyllead to 24-h nauplii of the brine shrimp *Artemia salina*, together with LC_{50} and LC_{100}.

Compound	0% effect	50% effect	100% effect
Tetramethyllead	180	250	670
Tetraethyllead	25	85	260

6.3 Toxicity to Fish

Appraisal

The toxicity of lead-contaminated water to fish varies considerably, depending on the availability and uptake of the lead ion. Factors affecting this availability are water hardness (presence of divalent anions), pH, salinity, and organic matter. Uptake is affected by the presence of other cations and the oxygen content of the water. Organic lead is taken up more readily than inorganic lead. The 96-h LC_{50} for inorganic lead in sensitive species can be as low as 1 mg dissolved lead/litre; nominal concentrations being up to 100 times higher in hard water. The few data available suggest that the toxicity of organic lead may be 10 to 100 times higher than that of inorganic lead. Long-term exposure of adult fish to inorganic lead induces sublethal effects on morphology, amino levulinic acid dehydratase (delta-ALAD) and other enzyme activities, and avoidance behaviour at available lead concentrations of 10-100 mg/litre. Juvenile stages are *generally more sensitive than adults, but eggs are often less sensitive because lead is adsorbed onto the egg surface and excluded from the embryo.*

The acute and subacute toxicity of lead to various species of fish and various life stages is summarized in Tables 5 and 6.

6.3.1 Toxicity of lead salts

Jones (1938) exposed stickleback (*Gasterosteus aculeatus*) to lead nitrate under static conditions, with the water replaced every 24 h, and observed the survival time over a range of doses. For adults 45-50 mm long, average survival times after exposure to 0.02, 0.5, and 20 mg lead nitrate/litre were 11 days, 81 h, and 6.5 h, respectively. For smaller adults (18-20 mm long) average survival times were 14 days, 10 days, and 2 days after exposure to 0.1, 0.5, and 3.0 mg/litre, respectively. The addition to the lead solutions (50 mg/litre) of calcium chloride at 2 mg/litre considerably lengthened the survival time; fish survived for more than 10 days, as long as the controls.

Davies et al. (1976) studied the acute toxicity of lead nitrate to rainbow trout (*Salmo gairdneri*) in 96-h static tests in hard and soft water. Lead salts tend to precipitate out in hard water and the authors' results were given in terms of both dissolved and total lead. For two bioassays in hard water, the 96-h LC_{50} values obtained were 1.32 and 1.47 mg dissolved lead nitrate/litre. The corresponding total lead values for the test water were 542 and 471 mg/litre, respectively. In a flow-through test using soft water, the 96-h LC_{50} was 1.17 mg/litre for both dissolved and total lead, since all of the salt was in solution. High levels of dissolved divalent anions in hard water, therefore, protect fish from lead by reducing its availability to them. This is also reflected in the results of Pickering & Henderson (1966), who conducted tests on a variety of fish species using lead chloride and lead acetate. There is a clear difference in their results between hard and soft water for the same species (Table 5). Results in this study are presented as total lead.

Lloyd (1961) pointed out that dissolved oxygen levels tend to be low in polluted water, while toxicity tests are conducted in water fully saturated with oxygen. He examined the effect of varying dissolved oxygen at low levels of lead salts, in that range of concentrations important for determining safe levels in water. At 65% oxygen, lead toxicity increased over that obtained using fully saturated water by a factor of 1.2 (ratio of concentrations which were equitoxic) and, at 40% saturation, by a factor of 1.45.

Davies et al. (1976) conducted long-term bioassays with rainbow trout to establish a maximum acceptable toxicant limit (MATC) for inorganic lead. The effects of lead nitrate on reproduction, egg survival, hatching success, and growth of the hatched larvae were assessed. In the first chronic test, fingerling trout were exposed to nominal total lead concentrations of 0, 40, 120, 360, 1080, or 3240 mg/litre. Actual dissolved lead was measured and results were

Table 5. Toxicity of lead salts to fish

Organisms	Life-stage/ size	Flow/ stat[a]	Temp. (°C)	Alkalinity[b]	Hardness[b]	pH	Salt	Parameter	Water concentration (mg/litre)	Reference
Fathead minnow (*Pimephales promelas*)	adult	stat	25	18	20	7.5	chloride	24-h LC50	8.18 (6.72-10.5)	Pickering & Henderson (1966)
	adult	stat	25	300	360	8.2	chloride	24-h LC50	482 (426-562)	
	adult	stat	25	18	20	7.5	chloride	48-h LC50	5.99 (4.31-8.69)	
	adult	stat	25	300	360	8.2	chloride	48-h LC50	482 (426-562)	
	adult	stat	25	18	20	7.5	chloride	96-h LC50	5.58 (3.94-7.89)	
	adult	stat	25	300	360	8.2	chloride	96-h LC50	482 (426-562)	
	adult	stat	25	18	20	7.5	acetate	24-h LC50	14.6 (10.7-63.9)	
	adult	stat	25	18	20	7.5	acetate	48-h LC50	10.4 (7.21-16.7)	
	adult	stat	25	18	20	7.5	acetate	96-h LC50	7.48 (4.86-11.8)	
Bluegill sunfish (*Lepomis macrochirus*)	adult	stat	25	18	20	7.5	chloride	24-h LC50	25.9 (22.5-30.4)	Pickering & Henderson (1966)
	adult	stat	25	300	360	8.2	chloride	24-h LC50	482 (426-562)	
	adult	stat	25	18	20	7.5	chloride	48-h LC50	24.5 (20.9-29.1)	
	adult	stat	25	300	360	8.2	chloride	48-h LC50	468 (410-549)	
	adult	stat	25	18	20	7.5	chloride	96-h LC50	23.8 (20.0-28.4)	
	adult	stat	25	300	360	8.2	chloride	96-h LC50	442 (379-524)	
Goldfish (*Carassius auratus*)	adult	stat	25	18	20	7.5	chloride	24-h LC50	45.4 (39.4-53.6)	Pickering & Henderson (1966)
	adult	stat	25	18	20	7.5	chloride	48-h LC50	31.5 (25.0-39.8)	
	adult	stat	25	18	20	7.5	chloride	96-h LC50	31.5 (25.0-39.8)	
	40-80 mm	stat	19-25		0	6.0-6.9	nitrate	48-h LC50	6.6 (4.7-9.2)	Weir & Hine (1970)
	40-80 mm	stat	19-25		50	6.0-6.9	nitrate	48-h LC50	110 (100-121)	Weir & Hine (1970)
Guppy (*Lebistes reticulatus*)	adult	stat	25	18	20	7.5	chloride	24-h LC50	24.5 (20.9-29.1)	Pickering & Henderson (1966)
	adult	stat	25	18	20	7.5	chloride	48-h LC50	24.5 (20.9-29.1)	
	adult	stat	25	18	20	7.5	chloride	96-h LC50	20.6 (16.4-26.8)	

Table 5 (contd).

Organisms	Life-stage/size	Flow/stat[a]	Temp. (°C)	Alkalinity[b]	Hardness[b]	pH	Salt	Parameter	Water concentration (mg/litre)	Reference
Bluegill sunfish (*Lepomis macrochirus*)		stat	20				nitrate	24-h LC50	6.3	Turnbull et al. (1954)
		stat	20				nitrate	48-h LC50	6.3	
Rainbow trout (*Salmo gairdneri*)	adult	flow	10.3-	86-94	133-137	7.7	nitrate	21-day LC50	2.3(1.6-3.3)	Hodson et al. (1978a)
	adult	stat	14	267	385	8.15	nitrate	96-h LC50	1.32 (measured; = 542 total lead)	Davies et al. (1976)
	adult	stat	10	30	32	6.85	nitrate	96-h LC50	1.17	Davies et al. (1976)
	adult	stat	7	29	30	6.85	nitrate	14-day LC50	0.20	
	juvenile	flow		82-132		6.4-8.3	nitrate	96-h LC50	8.0	Hale (1977)
Brook trout (*Salvelinus fontinalis*)	adult	flow	12	42.6	44.3	4.1	nitrate	96-h LC50	4.1	Holcombe et al. (1976)
Sarotherodon mossambicus		stat	28-32	7.7-11.7	32-38		nitrate	24-h LC50	> 40	Subbaiah et al. (1983)

Table 5 (contd).

Channel catfish	1.6 g	stat	18		44	7.1	arsenate	24-h LC50	> 100	Mayer &
(*Ictalurus*	1.6 g	stat	18		44	7.1	arsenate	96-h LC50	> 100	Ellersieck
punctatus)										(1986)
Mosquito fish	adult	stat	22-24	< 100		7.7-8.3	nitrate	24-h LC50	240	Wallen et al.
(*Gambusia*	adult	stat	22-24	< 100		7.7-8.3	nitrate	48-h LC50	240	(1957)
affinis)	adult	stat	22-24	< 100		7.7-8.3	nitrate	96-h LC50	240	
	adult	stat	18-20	< 100		7.1-7.2	oxide	24-h LC50	> 56 000	
	adult	stat	18-20	< 100		7.1-7.2	oxide	48-h LC50	> 56 000	
	adult	stat	18-20	< 100		7.1-7.2	oxide	96-h LC50	> 56 000	
Grey mullet	0.3-	flow	11-13	34.4-34.8[c]		6.9-8.5	nitrate	96-h LC50	> 4.5	Taylor et al.
(*Chelon labrosus*)	3.2 g									(1985)

[a] Stat = static conditions (water unchanged for duration of test); flow = flow-through conditions (lead concentration in water continuously maintained).
[b] Alkalinity and hardness expressed as mg/litre $CaCO_3$.
[c] These figures are values for salinity (expressed in $^{o}/_{oo}$), not alkalinity.

Table 6. Toxicity of organolead to fish

Organisms	Size (mm)	Flow/ stat[a]	Temp. (°C)	Alkalinity[b]	Hardness[b]	pH	Compound[d]	Parameter	Water concentration (mg/litre)	Reference
Bass (young) (*Morone labrax*)	6	stat	20				TML	48-h LC_{50}	0.10	Marchetti (1978)
	6	stat	20				TEL	48-h LC_{50}	0.065	
Tidewater silverside (*Menidia beryllina*)	40-100	stat	20		55	7.6-7.9	TML	96-h LC_{50}	13.5	Dawson et al. (1977)
Bluegill sunfish (*Lepomis macrochirus*)	33-75	stat	23		55	7.6-7.9	TML	96-h LC_{50}	84	Dawson et al. (1977)
	50-110	stat	20	33-81	84-163	6.9-7.5	TEL	24-h LC_{50}	2.0	Turnbull et al. (1954)
	50-110	stat	20	33-81	84-163	6.9-7.5	TEL	48-h LC_{50}	1.4	Turnbull et al. (1954)
							TEL	96-h LC_{50}	0.02	Wilber (1969)
Plaice (*Pleuronectes platessa*)	52	flow	15	34.9[c]			TML	96-h LC_{50}	0.05	Maddock &
	52	flow	15	34.9[c]			TEL	96-h LC_{50}	0.23	Taylor
	52	stat	15	34.9[c]			TriML	96-h LC_{50}	24.6	(1980)
	52	stat	15	34.9[c]			TriEL	96-h LC_{50}	1.7	Maddock &
	52	stat	15	34.9[c]			DML	96-h LC_{50}	300	Taylor
	52	stat	15	34.9[c]			DEL	96-h LC_{50}	75	(1980)

[a] Stat = static conditions (water unchanged for duration of test); flow = flow-through conditions (lead concentration in water continuously maintained).
[b] Alkalinity and hardness expressed as mg/litre $CaCO_3$.
[c] These figures are values for salinity (expressed in $^{o}/_{oo}$), not alkalinity.
[d] TML = tetramethyl lead; TEL = tetraethyl lead; TriML = trimethyl lead chloride; TriEL = triethyl lead chloride; DML = dimethyl lead dichloride; DEL = diethyl lead dichloride.

expressed in terms of dissolved salt. A MATC of between 0.018 and 0.032 mg/litre was found in terms of the "black tail effect". A similar bioassay with soft water suggested a MATC between 0.041 mg/litre, where no black tails occurred, and 0.076 mg/litre, where 4.7% of fish showed the black tail effect. These fish had been hatched from exposed eggs. When fingerlings from non-exposed eggs were used, in a soft water bioassay, the MATC for the black tail effect was between 0.072 mg/litre, when no black tails were seen, and 0.146 mg/litre, where 41.3% of fish had black tails. There were no significant differences between the measured dissolved lead concentrations in the two tests, indicating that fish from exposed eggs and sac fry were more sensitive to the effects of lead than those from non-exposed eggs. A long-term bioassay on reproductive effects established that reproductive females and eggs were relatively insensitive to lead. Therefore, the MATC is more realistic if based on the effects of lead on the sensitive fingerling stage. Brood fish in the reproductive test were exposed to lead concentrations, measured in the water, of 0.0005, 0.060, 0.077, 0.104, 0.175, and 0.270 mg/litre. Eggs and fry of the F_1 generation were exposed to measured lead at 0.0005, 0.060, 0.119, 0.238, 0.476, and 0.952 mg/litre. There was no mortality or effect on egg hatchability. The "black tail effect" was noted as the first stage of toxic symptoms to lead about 6 months after the hard-water study began. The entire caudal region at, or posterior to, the first caudal vertebra was blackened. Tail blackening of the tail was followed by spinal curvature and eroded caudal fins. This effect was noted in soft water tests at about 6 weeks. There was no effect in these studies on the growth of young trout, except where spinal curvature was so severe as to affect feeding.

Hodson et al. (1978a) similarly reported blackened tails in rainbow trout exposed to lead in the water at 0.120 mg/litre. After 32 weeks, 30% of the fish that survived showed black tails. The authors also reported that exposure to lead at concentrations as low as 0.013 mg/litre led to increases in red blood cell numbers, decreases in red blood cell volume, decreases in blood cell iron content, and decreases of red blood cell amino levulinic acid dehydratase activity (delta-ALAD). No changes in haematocrit or whole blood iron content were observed. The changes indicated increased production of red blood cells to compensate for increased death of red cells and inhibition of haemoglobin production. There was no significant uptake of lead from dietary dosing, though dietary lead might decrease uptake of dietary iron. All of the lead which causes toxic effects in fish is taken up directly from the water via the gills. Johansson-Sjobeck & Larsson (1979) also reported the depression of activity of delta-ALAD in rainbow trout exposed for 30 days to lead nitrate solutions (0.010, 0.075, and 0.30 mg/litre). The enzyme was depressed in red blood cells, spleen, and renal tissue. Fish exposed to the highest lead concentration also showed anaemia and basophilic stippling of the erythrocytes. White blood cells were not affected. Holcombe et al. (1976) exposed three generations of brook trout *(Salvelinus fontinalis*) to lead nitrate in the water. All second generation trout exposed to 0.235 or 0.474 mg total lead/litre, and 34% of those exposed to 0.119 mg/litre, developed spinal deformities. Scoliosis developed in 21% of newly hatched third generation fish exposed to 0.119 mg lead nitrate/litre. The weights of these same third generation fish were significantly reduced 12 weeks after hatching. The authors calculated a MATC for brook trout, based on the scoliosis effect, of between 0.058 and 0.119 mg total lead/litre (0.039 and 0.084 mg dissolved lead/litre) in soft water (hardness: 44 mg $CaCO_3$/litre) at a pH of between 6.8 and 7.6.

Hodson et al. (1980) examined the possibility that the toxic effect of lead on salmonids was due to ascorbic acid deficiency, since the symptoms were similar. They found no interaction between lead and ascorbic acid deficiency in their effects on the fish; thus, the toxicity of lead is not connected with ascorbic acid metabolism.

Weis & Weis (1977) exposed killifish eggs to lead nitrate at 0.1, 1.0, or 10 mg/litre in water. The lead was added at the start of the study and the solutions were not replaced. The authors stated that the removal of lead from the solution would have probably amounted to 79% over 96 h, the period of the test. The lead only slightly reduced axis formation in the embryos at all dose levels. At hatching, the fish were examined for malformations. Only 20% of the fry were normal when the added lead concentration was 1 mg/litre, the remainder having skeletal malformations. Some 40% of the fish could not uncurl from the position they had in the chorion and remained inactive, lying on the bottom of the dish. All fish exposed to 10 mg/litre lead were permanently curled. The curled fish could respond to tactile stimulation but returned to the curled position. Ozoh (1979) exposed eggs of the zebrafish (*Brachydario rerio*) to lead nitrate at 0, 0.036, or 0.072 mg/litre (measured) and monitored hatching success and abnormalities in the embryos. Compared with a control hatch rate of 80%, lead at 0.036 and 0.072 mg/litre gave rates of 19.8% and 27%, respectively. The presence of lead also resulted in poor absorption of yolk, erosion of the tail and fin, spinal curvature, and outgrowths from the fry (which appeared to be epitheliomas).

6.3.2 Biochemical effects

Christensen (1975) examined a range of biochemical parameters in brook trout embryos and alevins exposed to lead nitrate (0.057 mg/litre to 0.53 mg/litre) at the egg stage for 16 to 17 days, and then for a further 21 days as alevins. No effects on the eggs were seen. For alevins there was a decrease in weight, an increase in alkaline phosphatase activity, and an increase in acetylcholinesterase activity. Christensen et al. (1977) exposed brook trout (*Salvelinus fontinalis*) to lead nitrate at concentrations ranging from 0.009 mg/litre to 0.474 mg lead/litre for 2- or 8-week periods. They found no significant effects on body weight, body length, or on blood plasma glucose or lactic dehydrogenases. There were significant decreases in blood haemoglobin levels after exposure to lead at 58 μg/litre or more (after both 2 and 8 weeks). Plasma glutamic oxaloacetic transaminase activity was decreased after exposure to 34 μg/litre for both 2 and 8 weeks. Plasma sodium was elevated after exposure to 0.474 mg/litre for 8 weeks and chloride was elevated after exposure to 0.235 mg/litre for 2 weeks.

Hodson (1976) found that lead, as lead nitrate, in flowing water at concentrations as low as 13 μg/litre, caused a significant inhibition in the activity of red blood cell delta-ALAD in rainbow trout after 4 weeks. In a later study (Hodson et al., 1977), significant effects on delta-ALAD were observed within 2 weeks of exposure of rainbow trout, brook trout, goldfish, and pumpkinseed sunfish to lead concentrations of 10, 90, 470, and 90 μg/litre, respectively. The goldfish were affected by disease during the course of the test (4% mortality) and this result should be treated with caution. Jackim (1973) exposed mummichog (*Fundulus heteroclitus*) and winter flounder (*Pseudopleuronectes americanus*) to an initial concentration of 10 mg lead (as lead nitrate)/litre under static conditions in sea water. There were decreases of 22% and 18.5% in liver delta-ALAD activity in mummichogs after 96 h and 2 weeks, respectively. Winter flounder showed decreases of 66% and 58% in delta-ALAD activity in liver and kidney, respectively, after 1 week. It should be noted that the concentration of lead in solution at the end of the 2-week mummichog study was 0.8 mg/litre, only 8% of the initial concentration.

Shaffi (1979) examined various biochemical parameters in nine species of freshwater fish exposed to lead nitrate at nominal concentrations of 5, 10, 15, or 20 mg/litre. Lead caused glycogenolysis in all fish studied. The effect was greatest on muscle levels of carbohydrate, with lesser effects on liver, kidney, and brain. There was an inverse relationship between muscle, liver, and brain glycogen levels and the lead concentration in the water, and a direct relationship between lead levels in water and blood levels of glucose and lactose. Major carp were most affected, while various species of catfish and murrel were less sensitive to lead.

Sastry & Gupta (1978a) exposed catfish (*Channa punctata*) to lead nitrate at 3.8 mg/litre, previous tests having established that this concentration was sublethal. The fish were exposed for either 15 or 30 days and then sacrificed. Preparations were made of the stomach, intestine, pyloric caeca, and liver for the estimation of enzyme activities. There was no change in alkaline phosphatase activity in liver or stomach, but intestine and pyloric caeca enzyme activities showed marked inhibition after 15 days exposure. After 30 days exposure, alkaline phosphatase activity differed from the control level only in the pyloric caeca; now there was a marked elevation of activity. Alkaline phosphatase elevation is usually associated with cellular damage. After both 15 and 30 days of exposure, there was an elevation in acid phosphatase activity in all tissues. Three carbohydrases examined all showed an initial increase in activity followed by a marked decline. Proteases were elevated in activity throughout the experiment. In a later study, the same authors (Sastry & Gupta, 1978b) examined the effect of lead nitrate on digestive enzymes *in vitro*. There was a dose-related effect of lead, over the range 0.4, 0.8, and 1.6 μmol/litre, on the activities of alkaline phosphatase, lipase, tripeptide aminopeptidase, and glycylglycine dipeptidase. The inhibition caused by lead was reversed by the addition of EDTA.

Varanasi et al. (1975) reported effects on the properties of the epidermal mucus of rainbow trout exposed to lead chloride at concentrations between 0.1 and 1.0 mg/litre. Using electron spin resonance (ESR), the mucus was found to be more fluid after exposure to lead, and the effect persisted after removal of the lead from the water. The mucous characteristics of the epidermis affect swimming efficiency.

6.3.3 Behavioural effects

Giattina & Garton (1983) conducted avoidance behaviour experiments on rainbow trout using inorganic lead salts, and they concluded that trout will avoid lead at approximately 0.026 mg/litre (water hardness: 26 to 31 mg/litre). The value obtained by Jones (1948) of 0.4 mg/litre for the avoidance of lead in solution for the minnow *Phoxinus phoxinus* and the three-spined stickleback *Gasterosterus aculeatus* appears to be related to total lead rather than dissolved salt. No avoidance of total lead at 10, 20, or 40 mg/litre was found for green sunfish (*Lepomis cyanellus*) by Summerfelt & Lewis (1967). Weir & Hine (1970) pre-trained goldfish (*Carassius auratus*) to avoid electric shock with a light stimulus and then exposed them to solutions of lead nitrate. The lowest concentration of dissolved lead nitrate found to impair significantly the behavioural response was 0.07 mg/litre. The authors determined the lowest concentration of lead causing mortality in the same conditions to be 1.5 mg/litre. Addition of calcium carbonate to the test solution reduced the effect at higher lead concentrations. After exposure to lead nitrate at 10 mg/litre, the impairment of behavioural response was 70%. This was reduced to 25% by the addition of calcium carbonate at 50 mg/litre. The lead-exposed groups were retained and kept in clean water after the tests, and were re-tested for behavioural response at four weekly intervals. The effect of lead was permanent.

Ellgaard & Rudner (1982) exposed bluegill sunfish (*Lepomis macrochirus*) to concentrations of lead acetate ranging from 0.1 to 300 mg/litre. The LC_{50} for this species was found to be 400 mg/litre. Locomotor behaviour was monitored and no effects were noted. The absence of such effects at sublethal concentrations of metals is markedly unusual.

6.4 Toxicity to Amphibia

Appraisal

There is evidence that frog and toad eggs are sensitive to nominal lead concentrations of less than 1.0 mg/litre in standing water and 0.04 mg/litre in flow-through systems; arrested development and delayed hatching have been observed. For adult frogs, there are no significant effects below 5 mg/litre in aqueous solution, but lead in the diet at 10 mg/kg food has some biochemical effects.

Kaplan et al. (1967) exposed tree frogs (*Rana pipiens*) for 30 days to solutions of lead nitrate at between 25 and 300 mg lead/litre. They found sloughing of the integument, loss of postural tone, and sluggishness at all concentrations tested. All symptoms worsened with increasing lead concentration. Total red and white blood cell counts decreased progressively with increasing lead concentrations. Neutrophils and monocytes decreased at lower lead concentrations and all white cells at higher concentrations. The estimated LC_{50} was 105 mg/litre. Frogs exposed to lead nitrate at 500 mg/litre for 2 weeks, or 1000 mg/litre for 48 h, showed erosion of the gastric mucosa.

Khangarot et al. (1985) reported LC_{50} values for tadpoles of the frog *Rana hexadactyla* of 100, 66.7, 41.3, and 33.3 mg/litre after 24, 48, 72, and 96 h, respectively, at a temperature of 13-16 °C and a pH of 6.2-6.7.

Dilling & Healey (1926) exposed groups of one male and three female common frogs (*Rana temporaria*) for 3 weeks to water containing lead nitrate at concentrations between 16.5 and 3300 mg/litre. At the beginning of the study, the females were in full reproductive condition and gravid with eggs. The solutions were regularly changed, and pond weed was present in the tanks. All adult frogs died when exposed to concentrations of 330 mg/litre or more. At a lead nitrate concentration of 165 mg/litre, two batches of spawn were laid but no development of the embryos occurred. At 33 mg/litre, development commenced but proceeded no further than the late gastrula stage (day 10 of normal development). Only a few embryos developed when exposed to 16.5 mg/litre, and the tadpoles were 30% smaller than controls. Control animals produced two batches of spawn and all eggs developed. A further series of experiments, where only the spawn, and not the adults, was exposed to lead nitrate solutions, showed that lead affected development at concentrations much lower than those first tried. At 0.7 mg/litre, the development of most eggs was arrested, although those tadpoles which did develop were normal after a late hatch.

Birge et al. (1979) exposed narrow-mouth toad *(Gastrophryne carolinensis*) eggs to inorganic lead, in a continuous-flow bioassay, from fertilization through to 4 days post hatch (7 days exposure). They estimated an LC_{50} value of 0.04 mg/litre. Toads were found to be more sensitive to lead than goldfish or rainbow trout examined in parallel assays.

Ireland (1977) fed lead-contaminated earthworms to the African clawed toad (*Xenopus laevis*) for 8 weeks. The earthworm diet contained 10, 308, or 816 mg lead/kg. No toads died as a result of lead ingestion. There were no significant effects on growth rate, haemoglobin, haematocrit, or reticulocyte values, but blood delta-ALAD activity was significantly reduced.

7. TOXICITY TO TERRESTRIAL ORGANISMS

7.1 Toxicity to Plants

Appraisal

The tendency of inorganic lead to form highly insoluble salts and complexes with various anions, together with its tight binding to soils, drastically reduces its availability to terrestrial plants via the roots.

Translocation of the ion in plants is limited and most bound lead stays at root or leaf surfaces. As a result, in most experimental studies on lead toxicity, high lead concentrations in the range of 100 to 1000 mg/kg soil are needed to cause visible toxic effects on photosynthesis, growth, or other parameters. Thus, lead is only likely to affect plants at sites of very high environmental concentrations.

Bazzaz et al. (1974a) grew sunflowers (*Helianthus annuus*) plants in vermiculite in a controlled environment room. After 3 to 5 weeks, when the plants were 45 to 60 cm tall, the top 15 cm of each plant was excised and placed in a solution of lead salts (concentrations of 2, 20, 100, or 200 mg/litre) for 5 days. All doses caused a reduction in net photosynthesis and respiration over the exposure period. A 50% reduction in photosynthesis corresponded to a leaf tissue lead concentration of 193 mg/kg. In a second study, leaf peels were exposed to lead solutions ranging from 10 to 1000 μmol/litre, which caused reductions in stomatal opening of between 31% and 64%. The authors suggest that this effect accounts for the reduction of photosynthesis in the whole plant. Bazzaz et al. (1974b) grew corn and soybean plants in media. Nine days after germination, they were treated with lead chloride at concentrations varying between 250 and 4000 mg lead/litre. The photosynthetic rate, measured as carbon dioxide uptake, of leaves from corn plants was reduced to approximately 80% of the control level at concentrations of 500, 1000, or 2000 mg lead/litre, and was further reduced to 48% of the control level at 4000 mg lead/litre. The transpiration rate was reduced at all dose levels from approximately 67% of the control level at 250 mg lead/litre to an almost negligible rate at 4000 mg lead/litre. In soybeans, photosynthetic and transpiration rates were enhanced at 250 and 500 mg lead/litre. A reduction in the photosynthetic rate was found only at 4000 mg/litre, while transpiration was reduced at both 2000 and 4000 mg/litre.

Broyer et al. (1972) found no effect on the yield of commercial beans, barley, or tomato plants exposed to lead nitrate via a hydroponic culture solution at lead concentrations of up to 50 μg/litre.

Barker (1972) exposed explants of cauliflower inflorescence stem, lettuce stem, carrot root, and potato tubers to lead acetate at concentrations of between 0.005 and 50 mg/litre of medium over 20 days. There was a significant reduction in mean fresh weight of lettuce and carrot after exposure to lead concentrations of 0.005 mg/litre or more. Cauliflower and potato, both slower growing, showed significant reductions in yield only at 0.5 mg/litre or more.

Hooper (1937) studied the effect of lead sulfate in the hydroponic medium on the growth of dwarf French beans at concentrations ranging from 3 to 30 mg lead/litre. She adjusted the particular salts used in the medium to avoid the problem of lead salt precipitation. There was no effect on growth over a period of 1 month. Other species of plants were sprayed with a lead sulfate solution of 5 mg lead/litre. There was no effect on *Ulex europeus* or on *Lupinus arboreus*. Even spraying with supersaturated solutions which left a white coating of sulfate on the leaves had no appreciable effect.

Dilling (1926) exposed cress and mustard seeds to a solution of lead acetate (ranging from 0.5 to 5 g/litre, in terms of lead ion) for up to 25 days. A concentration of >0.5 g/litre delayed germination and initial growth. The delay increased with increasing lead concentration until, at 2.7 g/litre, only a few seeds germinated. At 5 g/litre, no germination occurred. Similar results were found when the author used lead nitrate solutions ranging from 0.01 to 10 g/litre, in terms of lead ion. Delayed germination and initial growth occurred at 0.1 g/litre or more, with no germination at 10 g/litre. The transfer of cress seeds to clean water after exposure to 0.7 or 1.5 g/litre for 18 days allowed germination and normal growth to take place.

Bell & Patterson (1926) started hyacinth bulbs over solutions of lead acetate from 0.0001 to 10 g/litre and found a graded inhibition of root growth over this concentration range. Bulbs developing in solutions of 1 or 10 g/litre showed complete arrest of root development and stunted flowers and leaves. The same bulbs showed stunting the following year when regrown over tap water.

Davis & Barnes (1973) dosed growing seedlings of loblolly pine (*Pinus taeda*) and red maple (*Acer rubrum*) with solutions of lead chloride between 2×10^{-4} and 5×10^{-3} mol/litre twice weekly for 2.5 months. Following exposure to 10^{-3} mol/litre or more, they observed a significant reduction in height and root dry weight for both species, and a reduction in stem dry weight for red maple. There was a significant reduction in pine stem dry weight at 5×10^{-3} mol/litre, and a significant increase in the maple leaf anthocyanin content at 10^{-3} mol/litre or more.

Keaton (1937) monitored the growth of pot-grown barley after the addition of lead nitrate or carbonate to the soil. At concentrations up to 3000 mg/kg soil, there were no deleterious effects on barley growth, and at low lead application rates, there was a small stimulation in barley growth (nitrate acts as a fertilizer at low rates). This stimulation was most marked at lead concentrations of between 0.1 and 0.4 mg/kg soil. Most of the lead was found to be fixed to the soil particles. Soluble lead available to the plant did increase with amount of salt added, but very little of the total lead was soluble. Oberlander & Roth (1978) measured the uptake of labelled potassium (^{42}K) into the roots and shoots of 7-day-old barley plants from nutrient solutions containing lead. Uptake was monitored over 5 h during exposure to lead at between 10^{-6} and 10^{-4} mol/litre. Potassium uptake was reduced significantly to 48% of the control level by a lead concentration of 10^{-4} mol/litre.

Dijkshoorn et al. (1979) added lead acetate, to give concentrations of between 11.4 and 1062 mg/kg, and fertilizer to sandy loam soil. The soil was placed into pots and three successive crops of plants were grown in the soil: plantain (*Plantago lanceolata*), clover (*Trifolium repens*), and ryegrass (*Lolium perenne*). Lead had no effect on plant yield even at the highest concentrations tested. The "uptake" of lead into the plant was at a constant ratio of 0.1 to the level in the soil. Lagerwerff et al. (1973) grew maize (*Zea mays)* and alfalfa in a greenhouse in silt loam at two soil pH levels (5.2 and 7.2), with lead chloride added to 64, 113, and 212 mg/kg. Total yield data (dry weight of plants) showed no effect of either lead or pH in maize. For alfalfa, there was no effect of lead at pH 5.2, but at pH 7.2 there was a significant increase in yield over controls with no lead. Baumhardt & Welch (1972) found that emergence, plant height, and grain yield of maize were not affected by a field application of lead acetate at a rate of 50 to 3200 kg/ha. No effects were noted on morphology, colour, maturity, or other growth parameters during the 2-year study. Carter & Wain (1964) investigated the use of lead nitrate as a fungicide in broad bean plants. The salt was toxic to fungi at sap concentrations of >0.1 mmol/litre, but was also toxic to the plant.

7.2 Toxicity to Invertebrates

Appraisal

Ingestion of lead-contaminated bacteria and fungi by nematodes leads to impaired reproduction. Woodlice seem unusually tolerant to lead, since prolonged exposure to soil or grass litter containing externally added lead salts had no effect. Caterpillars maintained on a diet containing lead salts show symptoms of toxicity leading to impaired development and reproduction.

The information available is too meagre to quantify the risks to invertebrates during the decomposition of lead-contaminated litter.

Doelman et al. (1984) incubated a mixed culture of bacteria in lead nitrate solutions and grew the fungus *Alternaria solani* on malt agar to which lead nitrate had been added. The cultures were used as food for the nematodes *Mesorhabditus monohystera* and *Aphelenchus avenae*, which were reared for up to 22 days on bacteria and fungus, respectively. Lead was taken up by bacteria to give a range of doses to the nematode of between 7.6 and 110 μg/g of food. All these exposures had a significant inhibitory effect on the reproduction of *Mesorhabditus monohystera*. A lead concentration of 2.47 μg/g in fungus strongly inhibited the reproduction of *Aphelenchus avenae* but variation was considerable; no statistics were presented for the fungal study.

Beyer & Anderson (1985) exposed woodlice (*Porcellio scaber*) to treated soil litter containing between 100 and 12 800 mg/kg dry weight of lead, as lead oxide, over 64 weeks. No significant effect was found on adult survival, number of young produced or on survival of young at exposures up to 6400 mg lead/kg. There was a significant reduction in all three parameters after exposure to 12 800 mg/kg. Beeby (1980) fed woodlice (*Porcellio scaber*) during "gestation" on cocksfoot grass (*Dactylis glomerata*) which had been dosed with lead (2911 or 16 483 mg/kg), as lead nitrate, and also on grass which had been collected from roadside verges. The verge grass contained 110 or 407 mg/kg (having been collected from two different sites). There was no deleterious effect at any of the exposure levels on the fertility of the woodlice after oviposition had occurred. Lead levels in gravid females correlated positively with body calcium levels and with the number of days on the contaminated diet.

Weismann & Skrobak (1980) fed the caterpillar *Scotia segetum* on a semisynthetic food to which lead had been added, and calculated LT_{50} values for lead chloride, at exposure levels of 250 and 500 mg/kg diet, of 72.1 and 28.7 h, respectively. For lead acetate at levels of 250 and 500 mg/kg diet, the LT_{50} values were 75.6 and 31.9 h, respectively. An increased ascorbic acid content in the diet (1000 mg/kg) reduced the lead toxicity by between 42% and 52%, but increased calcium in the diet (1000 mg calcium carbonate/kg) had no effect on lead toxicity. Weismann & Svatarakova (1981) fed the same species of caterpillar on natural diets contaminated with lead at various doses (50, 100, 200, 400, or 800 mg/kg diet) throughout development. There were reproductive effects at all doses, dependent on the instar of the larvae at first exposure. Only 20% of third instar larvae exposed to 50 mg lead/kg developed to the adult stage. These adults were deformed and the females failed to produce eggs. Only 40-73% of larvae fed on a diet containing lead at 50 to 200 mg/kg, from the third instar, produced pupae.

7.3 Toxicity to Birds

Appraisal

Lead salts are only toxic to birds at a high dietary dosage (100 mg/kg or more). Almost all of the experimental work is on chickens and other gallinaceous birds. Exposure of quail from hatching and up to reproductive age resulted in effects on egg production at dietary lead levels of 10 mg/kg. Although a variety of effects at high dosage have been reported, most can be explained as a primary effect on food consumption. Diarrhoea and lack of appetite, leading to anorexia and weight loss, are the primary effects of lead salts. Since there is no experimental evidence to assess effects on other bird species, it is necessary to assume a comparable sensitivity. If this is so, then it is highly improbable that environmental exposure would cause adverse effects.

Metallic lead is not toxic to birds except at very high dosage when administered in the form of powder. It is highly toxic to birds when given as lead shot; ingestion of a single pellet of lead shot can be fatal for some birds. The sensitivity varies between species and is dependent on diet. Since birds have been found in the wild with large numbers of lead shot in the gizzard (20 shot is not unusual), this poses a major hazard to those species feeding on river margins and in fields where many shot have accumulated.

There is little information on the effects of organolead compounds. Trialkyllead compounds produced effects on starlings dosed at 0.2 mg/day; 2 mg/day was invariably fatal.

The short-term and long-term dietary toxicity of lead salts and organolead is summarized in Table 7.

7.3.1 Toxicity of lead salts

Lead salts have low to moderate acute and short-term toxicity to birds. Lethal and severe sublethal effects have not been reported at levels likely to be found in the wild. Some sublethal effects have been noted after realistic exposure, but these are unlikely to affect bird populations.

7.3.1.1 Toxicity to birds' eggs

Ridgway & Karnofsky (1952) injected lead nitrate solutions into the yolk sac of chicken eggs, after 4 or 8 days of development, and into the chorio-allantoic membrane after 8 days of development. The LD_{50} was 0.30 at 4 days and 4.50 at 8 days, expressed as molar equivalents of lead, for the yolk sac route, and 3.00 molar equivalents at 8 days for the chorio-allantoic route. The 4-day result is equivalent to 0.10 mg lead nitrate/egg.

Haegele et al. (1974) dosed female mallards with 100 mg lead/kg diet. This was added as a mixture of 43 mg/kg lead carbonate, 37 mg/kg lead oxide, and 49 mg/kg lead sulfate, each salt contributing one-third of the total lead. No significant effect on eggshell thickness was found when it was measured on days 76 and 85 of treatment. When lead was added to the diet along with DDE at 40 mg/kg, lead did not increase the effect of the organochlorine on shell thickness.

7.3.1.2 Toxicity to adult and juvenile birds

Vengris & Mare (1974) exposed 6-week-old chickens to lead acetate in drinking water for 35 days at doses ranging from 20 to 640 mg lead/litre. The chickens were found to tolerate lead in the water at concentrations up to 160 mg/litre without showing any clinical or haematological signs, despite blood lead levels as high as

Table 7. Acute and dietary toxicity of lead to birds

Species	Age	Compound	Parameter	Concentration (mg/kg)	Reference
Japanese quail (*Coturnix coturnix japonica*)	3-4 months	tetraethyllead	acute LD_{50}[a]	24.6(14.7-41.3)	Hudson et al. (1984)
	14 days	powdered metallic lead	5-day LC_{50}	> 5000	Hill & Camardese (1986)[b]
	14 days	lead nitrate	5-day LC_{50}	> 5000	Hill & Camardese (1986)[b]
	14 days	lead sub-acetate	5-day LC_{50}	> 5000	Hill & Camardese (1986)[b]
	14 days	lead arsenate	5-day LC_{50}	2761(1622-4701)	Hill & Camardese (1986)[b]
Mallard duck (*Anas platyrhynchos*)	3-4 months	tetraethyllead	acute LD_{50}[a]	107(44.5-258)	Hudson et al. (1984)
	young	lead nitrate	< 100-day LC_{50}	> 500	DeWitt et al. (1963)
	adult	lead nitrate	< 100-day LC_{50}	> 50	DeWitt at al. (1963)

[a] Single oral dose expressed as mg/kg body weight.
[b] Hill & Camardese (1986) fed quail with a dosed diet for 5 days followed by a clean diet for 3 days.

6.2 mg/litre. At a dose of 320 mg lead/litre, the chickens exhibited early signs of lethargy and weakness, followed by anorexia, anaemia, and loss of weight. Peripheral paralysis occurred prior to death. Six out of twelve birds died within 30 days, and all surviving birds had decreased haemoglobin levels at 30 days. At the highest dose of 640 mg/litre, similar clinical signs were observed but all birds died within 34 days. Long-term exposure to lead at levels producing no clinical symptoms had no effect on antibody production against Newcastle disease virus.

In two separate studies, Damron et al. (1969) dosed 4-week-old broiler chickens with dietary lead acetate at levels between 10 and 2000 mg lead/kg for 4 weeks. They report that, at dietary lead levels of 100 mg/kg or less, there was no significant effect on body weight gain or on food consumption. At dosing levels of 1000 and 2000 mg/kg, there was a significant depression of body weight gain and food consumption.

Morgan et al. (1975) dosed newly-hatched Japanese quail with lead acetate in the diet at 10, 100, 500, and 1000 mg lead/kg for 5 weeks. There was a significant effect on body weight after dosing with 500 and 1000 mg lead/kg diet (food consumption was not monitored), and blood haemoglobin content was reduced in the same birds. A reduction in haematocrit was found after dosing with 1000 mg lead/kg diet, but only between weeks 4 and 5 of age. Relative weights of bursa, spleen, liver, and heart were not affected. After 5 weeks of dosing, testis size was reduced in birds fed 1000 mg lead/kg. All quail were able to express a normal primary humoral immune response following antigenic challenge with a saline suspension of sheep red blood cells, at 4 weeks of age. Relative adrenal weights were significantly increased after 5 weeks on the diets containing 500 or 1000 mg lead/kg. A similar experiment, but with the dosing beginning at 6 days of age, showed no adrenal effect.

Edens et. al. (1976) investigated the effects of dietary lead acetate on reproductive performance in Japanese quail *(Coturnix coturnix japonica)*. Chicks were reared from hatching on food to which lead acetate had been added to give 0, 1, 10, 100, or 1000 mg lead/kg. When chicks were 6 weeks old, they were transferred to a layer diet, similarly dosed with lead, and housed in pairs. The lighting schedule was continuous light for the first week, followed by 1 week on 10 h of light. Thereafter, lighting was increased by 1 h per day each week until the birds were receiving 14 h of light per day at 6 weeks of age. At this point they were paired. The quail were killed at 12 weeks of age. Records were kept of when females produced their first egg, rates of egg production, and hatchability of artificially-incubated eggs, together with body weights of adults. Only the highest dose rate (1000 mg/kg diet) affected growth of the birds. For the first 6 weeks, the body weight of treated birds, both males and females, was lower than that of controls. By the age of 12 weeks, treated males had caught up with controls but females were still significantly lighter. Egg production by females was depressed even at the lowest dose, and higher dose levels of lead acetate produced a greater effect. The highest dose level almost completely suppressed egg production and the few eggs produced at this dose level were soft-shelled or shell-less. Maximum rate of egg production was reached at 8 weeks of age in both control birds and females fed 1 mg/kg diet. This peak of egg production was delayed until the birds were 12 weeks old in the groups fed 10 or 100 mg/kg diet. There was also a delay in onset of egg laying, relative to controls, in groups fed 10, 100, and 1000 mg/kg diet. The highest dose also significantly delayed sexual maturity relative to other dosed groups. The hatch rate of eggs laid by groups fed 100 or 1000 mg/kg was significantly reduced.

Damron & Wilson (1975) conducted a series of studies to determine the toxicity of lead, in various forms, to bobwhite quail *(Colinus virginianus)*. At dietary dose rates of lead acetate of up to 1500 mg/kg during 6 weeks, juvenile birds showed no effect on body weight gain, food consumption, or mortality, and adult males showed no effect on semen quality or organ weights. Feeding birds at a dietary dose rate of 3000 mg/kg led to a significant depression in growth rate and an increase in mortality. In a similar study using white Chinese geese (Johnson & Damron, 1982), feeding lead acetate at dietary levels up to 2000 mg/kg had no effect on body weight or food consumption. At 2000 mg/kg diet, there was a slight increase in the size of the liver and some yellow discoloration.

Coburn et al. (1951) dosed adult mallard ducks *(Anas platyrhynchus)* daily with aqueous solutions of lead nitrate, introduced directly into the gizzard via a catheter. They found that a daily dose of 6 mg/kg body weight had no effect on body weight, red blood cell counts, or haemoglobin content over a period of 132 days, but with daily doses of 8 or 12 mg/kg body weight, there was a decrease in these parameters within 3 to 4 weeks. Kendall & Scanlon (1982) dosed adult male ringed turtle doves (*Streptopelia risoria*) with lead acetate, by intubation, at levels of 0, 25, 50, or 75 mg lead/kg body weight, daily for 7 days. At the highest rate of dosing, the birds lost 17% of their original body weight; weight loss was lower at the other two dosing rates (5% and 8% for 25 and 50 mg/kg per day, respectively). None of these weight changes was statistically significant. Schafer et al. (1983) estimated an 18-h LD_{50} for the red-winged blackbird (*Agelaius phoeniceus*) of >111 mg lead toxicity/kg body weight. This value was based on estimated intake from dosed food.

7.3.1.3 Enzyme effects

Dieter et al. (1976) established a correlation between the lead levels in the blood of canvasback ducks and the activity of the enzyme delta-ALAD. The ducks had taken up the lead from their natural environment. A level of 0.20 mg lead/litre blood was associated with a 75% decrease in enzyme activity. Kendall & Scanlon (1982) reported a similar correlation between lead residues in ring doves and delta-ALAD activity.

7.3.1.4 *Behavioural effects*

Frederick (1976) fed mallard ducklings on a diet containing lead nitrate (dissolved in propylene glycol) at 0, 5, 50, or 500 mg lead/kg diet. There was no effect of any of the treatments on the general activity of the ducklings after 3 and 8 days on these diets, but there was a significant, dose-related effect on weight gain.

Barthalmus et al. (1977) dosed trained pigeons by gastric intubation daily with 6.25, 12.5, or 25 mg lead acetate/kg body weight. The pigeons had been trained to peck response keys for a food reward in a complex system requiring multiple responses to obtain the reward. The lowest dose produced no significant effect on behavioural performance. The highest dose led to mortality after 18-35 days, and there were noticeable behavioural effects after 3-10 days. The middle dose of 12.5 mg/kg produced no deaths, but did significantly alter behavioural response after 30 days.

7.3.2 *Toxicity of metallic lead*

Lead shot taken into the gizzard of birds is highly toxic. Birds are affected or killed by small numbers of shot. Powdered lead appears to be less toxic, probably because it is not retained in the upper gut.

7.3.2.1 *Toxicity of powdered lead*

Hill & Camardese (1986) dosed Japanese quail with powdered metallic lead in the diet at doses ranging from 1000 to 5000 mg/kg diet. There was no mortality after 5 days on the lead-containing diet, or after a further 8 days of observation on a clean diet. At dose levels of 1495 or 2236 mg/kg diet, food consumption was unaffected.

Pattee (1984) fed American kestrels (*Falco sparverius*) with metallic lead in the diet at doses of 0, 10, or 50 mg/kg for 7 months. Although lead levels were elevated in the bones and liver of birds on treated diets, particularly at the highest dose level, no adverse effects were found with respect to survival, egg laying, initiation of incubation, fertility, or eggshell thickness. Hoffman et al. (1985a) dosed 1-day-old nestling American kestrels for 10 days with powdered metallic lead in corn oil daily (25, 125, or 625 mg/kg body weight per day). The birds were fed on mice in the mornings prior to dosing by intubation, and survivors were sacrificed on day 10. The only mortality occurred at the highest dose rate; 4 out of 10 birds died between days 6 and 8 of dosing. There was a significant effect on weight gain, but only at the two highest doses. After 10 days of dosing, birds given 625 mg/kg were 61% of control weight and birds given 125 mg/kg were 84% of control weight. Birds dosed at 25 mg/kg were 95% of control weight, not significantly different. In those groups which were affected, weight was reduced after days 4 and 5 of dosing. Mean brain weights of the groups given 625 and 125 mg/kg were

14% and 9%, respectively, lower than controls after 10 days. This reflected a general lack of growth because brain weight to body weight ratios were elevated relative to controls. There was also an effect on the skeleton, in addition to effects on soft tissues. Growth in both wing bones was reduced by 34-35% in the 625-mg/kg group and by 18-19% in the 125-mg/kg group. In a separate report (Hoffman et al., 1985b), the effects on biochemical and haematological indicators were given. Nestling American kestrels showed reduced haematocrit, haemoglobin level, and plasma creatine phosphorylase activity after 10 days of dosing with lead at 125 or 625 mg/kg body weight. Red blood cell delta-ALAD activity was depressed by these dose levels and also at 25 mg/kg. Brain, liver, and kidney delta-ALAD activities were inhibited by all lead treatments. Liver protein content and brain RNA to protein ratio decreased after lead treatment, whereas liver DNA, DNA to RNA ratio, and DNA to protein ratio increased. Brain monoamine oxidase and ATPase activity was not significantly altered by lead at these doses. The authors considered that these effects could explain, in part, the delayed development of the nestlings.

7.3.2.2 Toxicity of lead shot

Clemens et al. (1975) dosed adult mallard with five no. 6 lead shot and observed the birds over 20 days. The birds showed body weight loss over this period, together with clinical signs including green diarrhoea, anorexia, and weakness. High concentrations of lead in the blood, kidney, liver, and bone were recorded but there were lower concentrations in skeletal muscle. Birds on a high-fibre diet showed more severe clinical signs and higher tissue lead concentrations than birds on low-fibre diets. Mautino & Bell (1987) dosed mallard with two no. 4 lead shot and observed signs of lead toxicosis within 24 h. Varying degrees of paralysis, kinetic ataxia, or abnormal locomotor function were shown by 14 out of 17 birds. These neurological signs gradually disappeared and 8 days after dosing all birds appeared normal. The blood lead level was highest after 1 week at 7.8 mg/litre and remained significantly higher than the control value for a further 6 weeks. Blood samples were taken at weekly intervals. No lesions were found in the birds after 7 weeks. The effect of lead on blood delta-ALAD activity was maximal after 1 week, with 80% inhibition, and gradually returned to normal over the 7-week study.

Irwin & Karstad (1972) exposed adult mallard drakes for 14 weeks to concentrations of 17.8, 89, or 178 g of particulate lead/m^2 in a simulated marsh area. The mortality was 17%, 57%, and 100% for the three dose levels, respectively. All birds gave a positive fluorescent erythrocyte test and showed chronic lead toxicosis. Birds exposed to the highest concentration showed overt signs of lead poisoning and all died within 23 days. Finley et al. (1976) dosed male and female mallard with either one number 4 lead shot or one number 4 lead/iron combination shot (with 47% lead), and observed the birds for 4 weeks. No mortality was recorded and no tissue lesions were found. There was a correlation between lead residues in the bone and the number of eggs laid; the more eggs laid, the greater the residue of lead in the bone. This presumably reflects the greater movement of calcium out of bone to produce eggshells and its replacement from dietary calcium. After Dieter & Finley (1978) dosed male and female mallard with a single number 4 lead shot, two out of 60 birds died, showing signs typical of lead poisoning at necropsy. One month after dosing, the blood lead level was 0.317 mg/litre and the inhibition of erythrocyte delta-ALAD activity was 53%. After 3 months, inhibition was 30% and after 4 months was 15%, due to removal of lead from the circulation.

Chasko et al. (1984) captured wild mallard (*Anas platyrhynchos*) and black duck (*Anas rubripes*) and maintained them in captivity on a "natural diet" consisting of millet and buckweed, available at all times, together with duckweed, eelgrass, fish, sand shrimp, mussels, crabs, and snails, available for some of the time. Groups of 10 ducks (5 of each species) were dosed with 0, 2, or 5 lead shot or with 5 lead shot given singly over a 2-week period. More lead was accumulated in tissues from repeat dosing with single shot than with single dosing with 5 shot. Mortality was similar for the two species, with the black duck slightly more susceptible to lead. One out of 4 black ducks dosed with 2 shot died; 2 out of 4 died after dosing with 5 shot. Weight loss was also similar for both species. Birds with clear symptoms of lead poisoning showed a weight loss of about 20%; ducks which died had lost between 30% and 50% of body weight. Mortality generally increased with dose rate of lead shot. Grandy et al. (1968) dosed 15 mallard with 8 lead shot each and observed the effects over 30 days. Birds were also dosed with shot containing less lead (an alloy of 40% lead and 60% tin). Those mallard dosed with pure lead shot showed 100% mortality; all died between 5 and 15 days after dosing. Birds fed the alloy shot showed 27% mortality, with birds dying between 8 and 30 days after dosing. Rozman et al. (1974) dosed adult female mallard ducks with 8 lead shot, orally by gavage, and monitored serum enzyme activities over the next 14 days. They reported significant increases in the activity of serum glutamic pyruvic transaminase (SGPT) and decreases in that of serum alkaline phosphatase (SAP) after lead treatment. These enzyme changes were suspected to reflect tissue damage.

Chinese white geese dosed with a total of 200 lead shot over a 12-week period did not die (Johnson & Damron, 1982). This is in marked contrast to studies in other species where only a few shot caused death in a short time. Cook & Trainer (1966) exposed 10 Canada geese, some adult males, some females, and some immatures, to lead pellets (2-100 per bird) introduced directly into the oesophagus. The highest recorded blood lead level was 16.8 mg/litre in an immature bird dosed with 100 pellets. The lethal dose was found to be 4 to 5 pellets; the two birds dosed with 5 pellets died within 39 and 72 days, respectively. Regardless of the numbers of pellets introduced into the gizzard, there was uniform erosion of lead from the pellets. The rate of erosion of the pellets was initially very rapid, with a 65% to 70% loss of lead within the first 5 days. The pellets had almost disappeared within 35 days. Gross signs of lead toxicity included weakness and lethargy, anorexia, green diarrhoea, loss of weight, and oedematous heads. The loss of weight was most noticeable in birds given lower doses of lead, since those given high doses died while still retaining good body condition. Necropsy findings included impaction of the proventriculus, roughened and greenish staining of the gizzard lining, severe enteritis, distended gall bladder, discoloured liver, and flaccid heart. These pathological lesions were more noticeable in birds which survived longer and, therefore, had experienced the effects of the lead for longer periods.

Damron & Wilson (1975) found that dosing adult male bobwhite quail with 10 or more lead shot per week for 4 weeks increased mortality. More than 90% of males dosed with 30 lead shot per week died within 4 weeks.

Patee et al. (1981) dosed bald eagles with 10 lead shot each, repeating the dose if the bird succeeded in regurgitating the shot. Four out of five eagles died; the fifth was killed when it became blind 133 days after dosing. The time taken for the birds to die varied between 10 and 125 days, though three birds died within 20 days. Body weight loss varied between 16% and 23%, those birds which died quickly losing less weight than those surviving longer. In a study lasting 60 days, Stendell (1980) fed American kestrels (*Falco sparverius)* daily with either one number 9 shot (given in a dead mouse) or with mallard which had died from lead poisoning and contained residues of 27 to 34 μg/kg body weight. No kestrels died or exhibited visible signs of lead poisoning.

7.3.3 Toxicity of organolead compounds

Too few reports are available to demonstrate clearly the effects of organolead compounds on birds. It is of moderate toxicity to birds (Table 7). Tetraethyllead is readily converted to triethyllead in water and in animals. Results suggest that trialkyllead compounds are very toxic, but effects on only one species have been reported.

Haegele & Tucker (1974) showed that tetraethyllead had no effect on eggshell thickness in mallard ducks or Japanese quail at a dose of 6.0 mg/kg body weight over 6 days. There was a transitory effect, but normal thickness returned within the 6-day study.

Osborn et al. (1983) dosed starlings (*Sturnus vulgaris*) with either trimethyl- or triethyllead in two separate experiments at doses of 0, 0.2, or 2 mg/day for 11 days (approximately equivalent to 28 mg/kg body weight per day at the highest dose). All birds given the highest dose of trimethyl- or triethyllead died within 6 days. Predeath symptoms were relatively mild in the case of triethyllead, consisting of slightly slower respiratory rate and a tendency to squat, rather than stand, with fluffed-out feathers as if cold. The effects of trimethyllead were more dramatic. Within 24 h of the first dose, one of the birds was so badly coordinated that it was unable to perch or stand normally. Only a single bird, out of the group of six, appeared normal at this stage. Within 6 h of the second dose, all birds showed symptoms of lack of coordination. One was unable to place accurately its bill in the feeder. There was considerable weight loss. All birds died, or were killed for humanitarian reasons, within the first 5 days. Birds on the highest dose of trimethyllead (but not with triethyllead) had bright green watery droppings. Food consumption was greatly reduced at the high-dose levels; this was not surprising considering the lack of coordination. There was also an effect on the feeding behaviour of birds receiving 0.2 mg/day. They ate approximately the same amount of food, on average, as the control birds, but there was considerable variation from day to day in the amount eaten. This was noticeable after very few doses, possibly occurring after a single dose. Liver weights were significantly lower than those of controls in the case of the high dose of triethyllead and both the high and low doses of trimethyllead. Kidney weight was reduced only in birds receiving the high dose of trimethyllead.

7.4 Toxicity to Non-Laboratory Mammals

There are many reports of lead levels in wild mammals but few reports of toxic effects of the metal in the wild or in non-laboratory species.

Kilham et al. (1962) captured wild rats from the area of a dump at Hanover, New Hampshire, USA, which contained heavy metals. Nearly all of the sampled animals showed intranuclear inclusion bodies in the kidneys which were absent from populations of laboratory rats. These inclusions were identical in staining and electron-microscopic characteristics to similar bodies induced by lead in the laboratory. Renal tumours were found in some of the rats associated with the inclusion bodies. The livers of the trapped animals contained lead. Earlier reports (Hindle & Stevenson, 1930; Hindle, 1932; Syveston & Larson, 1947) showed similar inclusion bodies in rats trapped in sewers in London and New York. Zook et al. (1972) reported the killing of 34 simian primates and three fruit bats in Washington Zoo by lead in paint on their cages, and reviewed other examples of zoo animals poisoned by leaded paint.

8. EFFECTS OF LEAD IN THE FIELD

8.1 Tolerance of Plants to Lead

Plant tolerance to metals has been reviewed by Bradshaw et al. (1965), Antonovics et al. (1971), and Wainwright & Woolhouse (1975). Holl & Hampp (1975) and Peterson (1978) have reviewed the specific case of tolerance to lead. Most work has concentrated on plants growing on mining wastes rather than roadside verges.

The general conclusions are as follows. Metal tolerance is almost always specific, i.e., tolerance to one metal does not confer tolerance to others. There are degrees of tolerance, the metal content of particular soils correlating with the degree of tolerance of the local plant population. Tolerance is inherited, i.e., tolerant parents transmit tolerance to their offspring. Within a plant species, there are tolerant and sensitive populations. Tolerance, therefore, develops by selection, rather than by adaptation of individuals. Two possible mechanisms for tolerance, to metals in general, have been identified; an "external" mechanism prevents metal entering the plant, while an "internal" mechanisms allows entry but prevents the metal from coming into contact with sensitive processes within the organism.

Appraisal

Most work on plant tolerance to lead has concentrated on plants growing on mining wastes, naturally highly contaminated areas, and roadside verges. Tolerance has only been found in populations of a few plant species.

Jowett (1958) studied lead tolerance in the grasses *Agrostis tenuis* and *A. stolonifera* by measuring root growth in a culture solution containing lead nitrate at either 75 or 125 µmol/litre. Both species, from either "control areas" or from areas rich in metals other than lead, showed little tolerance. The growth, relative to plants not exposed to lead, was <37% and <22%, respectively, for the two lead levels. In *A. tenuis* from a mining area rich in zinc and lead, the values were 80% and 62%, respectively. Bradshaw (1952) grew *A. tenuis* taken from a disused lead mine (with 1% lead in the soil) and from an uncontaminated site 100 metres away. In uncontaminated soil, the plants from the mining area were smaller and grew more slowly than the plants from the contaminated area. In soil from the mine, plants collected from this site grew normally, whereas the others showed no growth (50% of tillers were dead or dying within 3 months).

Briggs (1972) collected the liverwort *Marchantia polymorpha* from city areas with soil lead concentrations of 252, 401, and 898 mg/kg dry weight and from a control area (28 mg/kg dry weight). The plants were exposed to lead nitrate in agar at a concentration of 400 mg lead/kg for 7 days and increase in thallus length was monitored. There was no effect on the plants from the city areas, but the control plant growth was significantly reduced.

Malone et al. (1974) showed that lead was concentrated in the cell walls of maize (*Zea mays*) and, therefore, excluded from interference with biochemical processes. Lead also tended to be concentrated on the surface of the roots of plants and excluded from the shoots.

8.2 Highways and Industrial Sources of Lead

Appraisal

No effect on the reproduction of birds nesting near highways has been observed. Toxic effects have been observed in pigeons in urban areas, the kidneys being most frequently affected.

In a report by Grue et al. (1984), swallows nesting near highways accumulated significant amounts of lead, but there were no effects on the number of eggs produced, number of nestlings, nestling body weights, or body weights of adults. In a similar study (Grue et al., 1986), starlings also accumulated lead but there were no effects on the same reproductive parameters. In feral pigeons (*Columba livia*) in London, Hutton (1980) detected effects including increased kidney weight, presence of renal inclusion bodies, altered kidney mitochondrial structure, and function and depression of delta-ALAD activity in blood, liver, and kidney. The effects were less than would have been predicted from laboratory experiments. The author suggested that factors, such as changes in the distribution of lead at the tissue and organelle level, and the antagonistic action of zinc, might be responsible.

Mierau & Favara (1975) measured lead in deer mouse populations close to roads, and considered that the residues were 5 times too low to cause any reproductive effects. Clark (1979) suggested that doses of lead ingested by little brown bats, shrews, and voles from roadside verges equalled or exceeded those which have caused mortality or reproductive impairment in domestic mammals. Lead concentrations in bats and shrews exceeded those concentrations found in mammals from mining areas showing renal abnormalities.

8.3 Lead Shot

Appraisal

Lead poisoning, due to the ingestion of lead shot, is a cause of death for large numbers of birds. In these cases, lead shot is found in the gizzards, and lead levels are elevated in the liver, kidneys, and bones.

A report from the Nature Conservancy Council's Working Group in the United Kingdom (NCC, 1981) discussed the problem of swan deaths attributable to lead poisoning. Mute swans in the United Kingdom showed 8% to 15% decreases in population numbers between 1955 and 1978. During the period 1961-1978, there were large differences in swan population changes in different parts of the country. Populations increased in northern Scotland, north Wales, and parts of eastern and southern England, whereas there were marked declines in central and southern Scotland, North-West England, the Midlands, South Wales, and the lower Thames Valley. Of the kills of swans reported between 1966 and 1978, 56% had no cause attributed, though some would have died from natural causes. In the years 1980 and 1981, the Ministry of Agriculture Fisheries and Food conducted postmortems on 288 mute swans. They reported that 39.2% of the swans had died from lead poisoning, the largest single cause of death found. Again there were regional differences with 50% of English swans dying of lead poisoning, but none of the Scottish swans. The source of the lead was either gun-shot or anglers split shot. The two can be distinguished using antimony content. Birds ingest particulate material, which may be contaminated with lead shot, to grind food in the gizzard before digestion. Postmortems on 299 mute swans carried out between 1973 and 1980 revealed gun-shot in only five birds. Other swan species are more likely to contain gun-shot; two-thirds of the Whooper and Bewick swans dying of lead poisoning on the Ouse Washes contained gun-shot. Lead from petrol in pleasure boats has been discounted as a source of lead in the birds. The report acknowledges the problem that lead use by anglers has not changed appreciably in 150 years, yet the elevated swan death rate is a recent phenomenon. The most likely explanation for this is the distribution of aquatic plant life. In recent years, marginal and submerged plants have been killed by pollution from boats and, more significantly, by the use of herbicides to keep channels clear. The lack of marginal plant life would make lead shot more available to swans. Other species of waterfowl also contain lead shot and are sometimes killed by it. These include greylag geese, mallard, pochard, tufted duck, and goldeneye. The highest incidence of lead shot contamination is in mallard in autumn at inland sites rather than coastal ones.

Gun-shot is a more important source of lead in birds in North America. Bagley et al. (1967) collected dead or dying Canada geese and found that dying birds showed marked cephalic oedema, with submandibular swellings, oedema of eyelids, and a profuse discharge from eyes and nares. Shot was found in the gizzards of the geese and high lead levels were recorded in liver, tibia, and kidney. Anderson (1975) studied about 1500 waterfowl dying at Rice Lake, Illinois, USA, and found that lesser scaup made up 75% of 394 birds collected dead or dying. Of 96 scaup examined, 75% had lead, at least one pellet, in the gizzard. Lead levels averaged 46 mg/kg in the liver and 66 and 40 mg/kg in kidney and wing bone, respectively. The incident occurred following a period of drought which killed food plants. With a return to normal water levels, plants began to grow again but lead pellets were more readily available in the feeding sites.

Trainer & Hunt (1965) estimated that 1700 Canada geese succumbed to lead poisoning in Wisconsin between 1940 and 1965. Other species were also affected. Lewis & Ledger (1968) found that mourning doves taken from a public field managed for shooting contained lead shot. Of 1949 gizzards examined, 1% contained between 1 and 24 shot. Examination of the area revealed 10 890 pre-shooting and 43 560 post-shooting shot per acre. Locke & Bagley (1967) found that gizzards from 4 out of 62 shot birds contained lead and that lead levels in 43 livers ranged from 0.4 to 14 mg/kg.

8.4 Organic Lead

Appraisal

A recurring incident of massive bird kills in estuaries near to industrial plants manufacturing leaded "anti-knock" compounds has been reported. The total lead content of the livers was sufficiently high to cause mortalities: lead was mostly present in the alkyl form.

In the autumn of 1979, about 2400 birds were found dead or dying in the Mersey estuary, United Kingdom, the majority being dunlin, a wader (Bull et al., 1983). Smaller numbers were found in 1980 and 1981. There is a plant manufacturing petrol additives in the vicinity. Affected birds contained elevated lead levels, mostly as alkyllead. The livers of dead birds from the incident contained an average of 11.14 mg total lead/kg wet weight, sick birds 8.85 mg/kg, apparently healthy birds from the same area 4.5 mg/kg, and healthy birds from another estuary 0.14 mg/kg. The authors note that Head et al. (1980) found 1 mg lead/kg in *Macoma balthica*, a food source for the waders, during the incident. Bull et al. (1983) concluded that total liver lead was sufficiently high to result in death. It was mostly in the form of alkyllead, which is at least as toxic as inorganic lead (Osborn et al., 1983). Symptoms were similar to those of inorganic lead poisoning and dissimilar to the effects of other pollutants present in the area. The high liver concentration, compared to the kidney concentration, was taken to indicate a recent acute exposure. There were no other toxic chemicals in significant amounts in the area, and there was no indication of disease. In the area discharging waste to the estuary, there was an industrial source manufacturing anti-knock compounds.

Gill et al. (1960) investigated effluent output from tetraethyllead production plants to assess the likely environmental hazard of a new plant. They measured 48-h LC_{50s} of the effluent, containing some alkyllead, for three-spined stickleback and coho salmon at 14 g/litre, and they concluded that the effluent would pose no hazard. No attempt was made to assess indirect hazard caused to birds by food organisms concentrating the lead.

In 1974, the 2000 ton cargo ship, "Cavtat", sank in a water depth of 94 m, 5.6 km from the Adriatic coast of Italy. Its cargo consisted of 325 tons of lead anti-knock compounds. At the time of recovery of the vessel, a loss of 7% of this cargo was estimated. Tiravanti & Boari (1979) concluded that the lead compounds were restricted to a limited area around the wreck and, based on water concentrations of <10 μg alkyllead/litre, had no significant environmental effect.

9. EVALUATION

9.1 General Considerations

In evaluating the environmental hazard of lead, it is necessary to extrapolate from laboratory studies to ecosystems. This must be done with extreme caution for the following reasons.

(a) The availability of lead to organisms in the environment is limited by its strong adsorption to environmental components, such as soil sediment, organic matter, and biota. It is accepted that biomagnification of lead does not take place; i.e., there is no increase in concentration of the metal in food-chains. However, environmental contamination with lead is widespread and organisms do accumulate high body burdens of lead.

(b) Environmental variables such as temperature, pH and chemical composition of water, soil type, and geology have been shown in limited studies on a narrow range of species to affect both the uptake and the effect of lead.

(c) Available, rather than nominal or total, lead is the determinant parameter in assessing uptake by, and effects on, organisms.

(d) There is limited data from controlled experimental studies on the effects of mixtures of metals. Organisms in the environment are exposed to mixtures of pollutants. Acid deposition can release various metals into the environment.

(e) Little experimental work has been carried out on species or communities that are either representative or key components of natural communities and ecosystems. Studies have not considered all of the interactions between populations and all of the environmental factors affecting these populations.

It is probable that subtle disturbances to the community would occur at much lower concentrations than those suggested in laboratory studies on acute effects. Much of the available information on lead toxicity is based on experimental studies carried out at unrealistically high nominal concentrations and short-term exposure. This makes it difficult to extrapolate to field conditions.

9.2 The Aquatic Environment

Lead enters the aquatic environment through surface runoff and deposition of airborne lead. Adsorption to sediments occurs rapidly and almost quantitatively.

The uptake and accumulation of lead by aquatic organisms from water and sediments are influenced by various environmental factors. These must be taken into consideration when evaluating the hazards of environmental contamination by lead.

Lead uptake by aquatic organisms is slow and reaches equilibrium only after prolonged exposure. Aquatic organisms at low trophic levels show a much higher accumulation of lead than those at higher trophic levels, reaching bioconcentration factors of up to 100 000. On the other hand, biomagnification through food chains is very low, often exhibiting values far below 1. However, this by no means indicates the absence of hazard.

The toxicity of lead to aquatic organisms varies considerably depending on availability, uptake, and species sensitivity; generally, the earlier life stages are more vulnerable. Lead interferes with biochemical, physiological, morphological, and behavioural parameters.

Organolead compounds are generally 10-100 times more toxic to aquatic organisms than is inorganic lead. Tetraalkyllead becomes toxic by conversion into trialkyllead.

9.3 The Terrestrial Environment

Lead is introduced to terrestrial communities by atmospheric deposition on to exposed surfaces. There is insufficient evidence to indicate a hazard to terrestrial organisms from airborne lead. Normal concentrations of lead in soil range from 15 to 30 mg/kg; roadside soils can reach 5000 mg/kg and soils from industrial sites may exceed 30 000 mg/kg. Although soil retards the movement of lead through terrestrial communities, some lead may be leached from highly contaminated soils. Some soil lead is taken up by plants and passed to animals, but a major fraction is accumulated at the surface of root cells. Some of the factors that determine availability to plants are pH, organic matter, and soil type. Generally, lead is not toxic to plants at soil concentrations below 1000 mg/kg. Some plant populations can tolerate higher concentrations, and some appear to develop a genetic tolerance. Animals are exposed to lead through the ingestion of water, food, soil, and dust. In all cases, the concentrations in animals are related to environmental concentrations, and in most cases, lead appears to accumulate preferentially in calcified tissues. Certain bird populations are also exposed to lead shot.

It is improbable that environmental exposures cause acute adverse effects in most terrestrial populations. However, lead shot is a major hazard in certain bird populations that tend to ingest gravel into the gizzard to grind food. Laboratory studies indicate that the expected effects on animals would be changes in behaviour, disruption of haematological metabolism, and inhibition of certain enzymes. There may be a strong correlation with calcium metabolism.

REFERENCES

AICKIN, R.M. & DEAN, A.C.R. (1978) Lead accumulation by microorganisms. *Microbios Lett.*, **5**: 129-134.

ANDERSON, R.V. (1978) The effects of lead on oxygen uptake in the crayfish, *Orconectes virilis* (Hagen). *Bull. environ. Contam. Toxicol.*, **20**: 394-400.

ANDERSON, W.L. (1975) Lead poisoning in waterfowl at Rice lake, Illinois. *J. wildl. Manage.*, **39**: 264-270.

ANDERSON, R.L., WALBRIDGE, C.T., & FIANDT, J.T. (1980) Survival and growth of *Tanytarsus dissimilis* (Chironomidae) exposed to copper, cadmium, zinc and lead. *Arch. environ. Contam. Toxicol.*, **9**: 329-335.

ANTONOVICS, J., BRADSHAW, A.D., & TURNER, R.G. (1971) Heavy metal tolerance in plants. *Adv. ecol. Res.*, **7**: 1-85.

APOSTOL, S. (1973) A bioassay of toxicity using protozoa in the study of aquatic environment pollution and its prevention. *Environ. Res.*, **6**: 365-372.

ASH, C.P.L. & LEE, D.L. (1980) Lead, cadmium, copper and iron in earthworms from roadside sites. *Environ. Pollut.*, **22**: 59-67.

AYLING, G.M. (1974) Uptake of cadmium, zinc, copper, lead and chromium in the Pacific oyster *Crassostrea gigas* grown in the Tamar River, Tasmania. *Water Res.*, **8**: 729.

BABICH, H. & STOTZKY, G. (1979) Abiotic factors affecting the toxicity of lead to fungi. *Appl. environ. Microbiol.*, **38**: 506-513.

BABICH, H. & STOTZKY, G. (1983) Influence of chemical speciation on the toxicity of heavy metals to the microbiota. In: Nriagu, J.O., ed. *Aquatic toxicology,* New York, Chichester, Brisbane, Toronto, John Wiley & Sons, pp. 1-46.

BAGLEY, G.E. & LOCKE, L.N. (1967) The occurrence of lead in tissues of wild birds. *Bull. environ. Contam. Toxicol.*, **2**: 297-305.

BAGLEY, G.E., LOCKE, L.N., & NIGHTINGALE, G.T. (1967) Lead poisoning in Canada geese in Delaware. *Avian Dis.*, **11**: 601-608.

BARKER, W.G. (1972) Toxicity levels of mercury, lead, copper, and zinc in tissue culture systems of cauliflower, lettuce, potato, and carrot. *Can. J. Bot.*, **50**: 973-976.

BARTHALMUS, G.T., LEANDER, J.D., MCMILLAN, D.E., MUSHAK, P., & KRIGMAN, M.R. (1977) Chronic effects of lead on schedule-controlled pigeon behavior. *Toxicol. appl. Pharmacol.*, **42**: 271-284.

BAUDOUIN, M.F. & SCOPPA, P. (1974) Acute toxicity of various metals to freshwater zooplankton. *Bull. environ. Contam. Toxicol.*, **12**: 745-751.

BAUMHARDT, G.R. & WELCH, L.F. (1972) Lead uptake and corn growth with soil-applied lead. *J. environ. Qual.*, **1**: 92-94.

BAZZAZ, F.A., CARLSON, R.W., & ROLFE, G.L. (1974a) The effect of heavy metals on plants: Part I. Inhibition of gas exchange in sunflower by Pb, Cd, Ni and Tl. *Environ. Pollut.*, **7**: 241-246.

BAZZAZ, F.A., ROLFE, G.L., & WINDLE, P. (1974b) Differing sensitivity of corn and soybean photosynthesis and transpiration to lead contamination. *J. environ. Qual.*, **3**: 156-158.

BEEBY, A. (1980) Lead assimilation and brood-size in the woodlouse *Porcellio scaber* Crustacea, Isopoda following oviposition. *Pedobiologia*, **20**: 360-365.

BELL, W.B. & PATTERSON, J. (1926) The effect of metallic ions on the growth of hyacinths. *Ann. appl. Biol.*, **13**: 157-159.

BENGTSSON, G. & RUNDGREN, S. (1984) Ground-living invertebrates in metal-polluted forest soils. *Ambio*, **13**: 29-33.

BEYER, W.N. (1986) A reexamination of biomagnification of metals in terrestrial food chains. *Environ. Toxicol. Chem.*, **5**: 863-864.

BEYER, W.N. & ANDERSON, A. (1985) Toxicity to woodlice of zinc and lead oxides added to soil litter. *Ambio*, **14**: 173-174.

BEYER, W.N. & MOORE, J. (1980) Lead residues in eastern tent caterpillars *(Malacosoma americanum)* and their host plant *(Prunus serotina)* close to a major highway. *Environ. Entomol.*, **9**(1): 10-12.

BEYER, W.N., CHANEY, R.L., & MULHERN, B.M. (1982) Heavy metal concentrations in earthworms from soil amended with sewage sludge. *J. environ. Qual.*, **11**: 381-385.

BIESINGER, K.E. & CHRISTENSEN, G.M. (1972) Effects of various metals on the survival, growth, reproduction, and metabolism of *Daphnia magna*. *J. Fish. Res. Board Can.*, **29**: 1691-1700.

BIRDSALL, C.W., GRUE, C.E., & ANDERSON, A. (1986) Lead concentrations in bullfrog *Rana catesbeiana* and green frog *Rana clamitans* tadpoles inhabiting highway drains. *Environ. Pollut.*, **40**: 233-247.

BIRGE, W.J., BLACK, J.A., & WESTERMAN, A.G. (1979) Evaluation of aquatic pollutants using fish and amphibian eggs as bioassay organisms. In: *Animals as monitors of environmental pollutants*, Washington, DC, National Academy of Sciences, pp. 108-118.

BORGMANN, U., KRAMAR, O., & LOVERIDGE, C. (1978) Rates of mortality, growth, and biomass production of *Lymnaea palustris* during chronic exposure to lead. *J. Fish. Res. Board Can.*, **35**: 1109-1115.

BOUTET, C. & CHAISEMARTIN, C. (1973) Propriétés toxiques spécifiques des sels métalliques chez *Austropotamobius pallipes pallipes* et *Orconectes limosus*. *C. R. Soc. Biol. (Paris)*, **167**: 1933-1938.

BRADSHAW, A.D. (1952) Populations of *Agrostis tenuis* resistant to lead and zinc poisoning. *Nature (Lond.)*, **169**: 1098.

BRADSHAW, A.D., MCNEILLY, T.S., & GREGORY, R.P.G. (1965) Industrialization, evolution and the development of heavy metal tolerance in plants. In: Goodman, G.T., Edwards, R.W., & Lambert, J.M., ed. *Ecology and the industrial society*, Oxford, Blackwell Scientific Publications, pp. 327-343 (British Ecology Society Symposium 5).

BRIGGS, D. (1972) Population differentiation in *Marchantia polymorpha* L. in various lead pollution levels. *Nature (Lond.)*, **238**: 166-167.

BRINGMANN, G. & KUHN, R. (1959a) The toxic effects of waste water on aquatic bacteria, algae and small crustaceans. *Gesund.-Ing.*, **80**: 115.

BRINGMANN, G. & KUHN, R. (1959b) Water toxicology studies with protozoans as test organisms. *Gesund.-Ing.*, **80**: 239.

BROWN, B. & AHSANULLAH, M. (1971) Effects of heavy metals on mortality and growth. *Mar. Pollut. Bull.*, **2**: 182-187.

BROWN, B.T. & RATTIGAN, B.M. (1979) Toxicity of soluble copper and other metal ions to *Elodea canadensis*. *Environ. Pollut.*, **20**: 303-314.

BROYER, T.C., JOHNSON, C.M., & PAULL, R.E. (1972) Some aspects of lead in plant nutrition. *Plant Soil*, **36**: 301-313.

BUGGIANI, S.S. & RINDI, S. (1980) Lead toxicosis and salt glands in domestic ducks. *Bull. environ. Contam. Toxicol.*, **24**: 152-155.

BULL, K.R., EVERY, W.J., FREESTONE, P., HALL, J.R., OSBORN, D., COOKE, A.S., & STOWE, T. (1983) Alkyl lead pollution and bird mortalities on the Mersey estuary, U.K., 1979-1981. *Environ. Pollut.*, **31**: 239-259.

CALABRESE, A. & NELSON, D.A. (1974) Inhibition of embryonic development of the hard clam *Mercenaria mercenaria* by heavy metals. *Bull. environ. Contam. Toxicol.*, **11**: 92-97.

CALABRESE, A., COLLIER, R.S., NELSON, D.A., & MCINNES, J.R. (1973) The toxicity of heavy metals to embryos of the American oyster *Crassostrea virginica*. *Mar. Biol.*, **18**: 162-166.

CANNON, H.L. & BOWLES, J.M. (1962) Contamination of vegetation by tetraethyl lead. *Science*, **137**: 765-766.

CARTER, G.A. & WAIN, R.L. (1964) Investigations on fungicides. XI. The fungitoxicity, phytotoxicity, and systemic fungicidal activity of some inorganic salts. *Ann. appl. Biol.*, **53**: 291-309.

CHASKO, G.G., HOEHN, T.R., & HOWELL-HELLER, P. (1984) Toxicity of lead shot to wild black ducks and mallards fed natural foods. *Bull. environ. Contam. Toxicol.*, **32**: 417-428.

CHINNAYYA, B. (1971) Effect of heavy metals on the oxygen consumption by the shrimp *Caridina rajadhari Bouvier*. *Indian J. exp. Biol.*, **9**: 277-278.

CHOW, T.J. (1970) Lead accumulation in roadside soil and grass. *Nature (Lond.)*, **225**: 295-296.

CHRISTENSEN, E.R., SCHERFIG, J., & DIXON, P.S. (1979) Effects of manganese, copper and lead on *Selenastrum capricornutum*. *Water Res.*, **13**: 79-92.

CHRISTENSEN, G.M. (1975) Biochemical effects of methylmercuric chloride, cadmium chloride, and lead nitrate on embryos and alevins of the brook trout, *Salvelinus fontinalis*. *Toxicol. appl. Pharmacol,*, **32**: 191-197.

CHRISTENSEN, G., HUNT, E., & FIANDT, J. (1977) The effect of methylmercuric chloride, cadmium chloride and lead nitrate on six biochemical factors of the brook trout *Salvelinus fontinalis*. *Toxicol. appl. Pharmacol.*, **42**: 523-530.

CLARK, D.R. (1979) Lead concentrations: bats versus terrestrial small mammals collected near a major highway. *Environ. Sci. Technol.*, **13**: 338-341.

CLELAND, K.W. (1953) Heavy metals, fertilization and cleavage in the eggs of *Psammechinus miliaris*. *Exp. cell Res.*, **41**: 246-248.

CLEMENS, E.T., KROOK, L., ARONSON, A.L., & STEVENS, C.E. (1975) Pathogenesis of lead shot poisoning in the mallard duck. *Cornell Vet.*, **65**: 248-285.

CLOUTIER, N.R., CLULOW, F.V., LIM, T.P., & DAVE, N.K. (1986) Metal (Cu, Ni, Fe, Co, Zn, Pb) and Ra-226 levels in tissues of meadow voles *Microtus pennsylvanicus* living on nickel and uranium mine tailings in Ontario, Canada: site, sex, age and season effects with calculation of average skeletal radiation dose. *Environ. Pollut.*, **41**: 295-314.

COBURN, D.R., METZLER, D.W., & TREICHLER, R. (1951) A study of absorption and retention of lead in wild waterfowl in relation to clinical evidence of lead poisoning. *J. wildl. Manage.*, **15**: 186-192.

COOK, R.S. & TRAINER, D.O. (1966) Experimental lead poisoning of Canada geese. *J. wildl. Manage.*, **30**: 1-8.

COOMBS, T.L. (1977) Measurement and toxicity of metallic and organic species. *Proc. Anal. Div. Chem. Soc.*, **14**: 219-222.

CRIST, T.O., WILLIAMS, N.R., AMTHOR, J.S., & SICCAMA, T.G. (1985) The lack of an effect of lead and acidity on leaf decomposition in laboratory microcosms. *Environ. Pollut.*, **38**: 295-303.

DAMRON, B.L., SIMPSON, C.F., & HARMS, R.H. (1969) The effect of feeding various levels of lead on the performance of broilers. *Poult. Sci.*, **48**: 1507-1509.

DAMRON, B.L. & WILSON, H.R. (1975) Lead toxicity of bobwhite quail. *Bull. environ. Contam. Toxicol.*, **14**: 489-496.

DAVIES, P.H., GOETTL, J.P., SINLEY, J.R., & SMITH, N.F. (1976) Acute and chronic toxicity of lead to rainbow trout *Salmo gairdneri*, in hard and soft water. *Water Res.*, **10**: 199-206.

DAVIS, J.B. & BARNES, R.L. (1973) Effects of soil-applied fluoride and lead on growth of loblolly pine and red maple. *Environ. Pollut.*, **5**: 35-44.

DAWSON, G.W., JENNINGS, A.L., DROZDOWSKI, D., & RIDER, E. (1977) The acute toxicity of 47 industrial chemicals to fresh and saltwater fish. *J. hazard. Mater.*, **1**: 303-318.

DERMOTT, R.M. & LUM, K.R. (1986) Metal concentrations in the annual shell layers of the bivalve *Elliptio complanata*. *Environ. Pollut.*, **12**: 131-143.

DEWITT, J.B., STICKEL, W.H., & SPRINGER, P.F. (1963) Wildlife studies, Patuxent Wildlife Research Center. In: *Pesticide-Wildlife Studies. A review of Fish and Wildlife Service investigations during 1961 and 1962*, Washington, DC, US Department of the Interior, Fish and Wildlife Service, pp. 71-96 (Circular No. 167).

DIETER, M.P. & FINLEY, M.T. (1978) Erythrocyte delta-aminolevulinic acid dehydrogenase activity in mallard ducks: duration of inhibition after lead shot dosage. *J. wildl. Manage.*, **42**: 621-625.

DIETER, M.P., PERRY, M.C., & MULHERN, B.M. (1976) Lead and PCBs in canvasback ducks: relationship between enzyme levels and residues in blood. *Arch. environ. Contam. Toxicol.*, **5**: 1-13.

DIJKSHOORN, W., VAN BROEKHOVEN, L.W., & LAMPE, J.E.M. (1979) Phytotoxicity of zinc, nickel, cadmium, lead, copper, and chromium in three pasture plant species supplied with graduated amounts from the soil. *Neth. J. agric. Sci.*, **27**: 241-253.

DILLING, W.J. (1926) Influence of lead and the metallic ions of copper, zinc, thorium, beryllium, and thallium on the germination of seeds. *Ann. appl. Biol.*, **13**: 160-167.

DILLING, W.J. & HEALEY, C.W. (1926) Influence of lead and the metallic ions of copper, zinc, thorium, beryllium and thallium on the germination of frogs spawn and on the growth of tadpoles. *Ann. appl. Biol.*, **13**: 177-188.

DOELMAN, P., NIEBOER, G., SCHROOTEN, J., & VISSER, M. (1984) Antagonistic and synergistic toxic effects of Pb and Cd in a simple food chain: nematodes feeding on bacteria or fungi. *Bull. environ. Contam. Toxicol.*, **32**: 717-723.

DOLLARD, G.J. (1986) Glasshouse experiments on the uptake of foliar applied lead. *Environ. Pollut.*, **40**: 109-119.

EDELMAN, W.Ma.Th., VAN BEERSUM, I., & JANS, Th. (1983) Uptake of cadmium, zinc, lead, and copper by earthworms near a zinc-smelting complex: Influence of soil pH and organic matter. *Bull. environ. Contam. Toxicol.*, **30**: 424-427.

EDENS, F.W., BENTON, E., BURSIAN, S.J., & MORGAN, G.W. (1976) Effect of dietary lead on reproductive performance in Japanese quail *Coturnix coturnix japonica*. *Toxicol. appl. Pharmacol.*, **38**: 307-314.

EISLER, R. (1977) Acute toxicities of selected heavy metals to the softshell clam, *Mya arenaria*. *Bull. environ. Contam. Toxicol.*, **17**: 137-145.

ELLGAARD, E.G. & RUDNER, T.W. (1982) Lead acetate: toxicity without effects on the locomotor activity of the bluegill sunfish, *Lepomis macrochirus* Rafinesque. *J. Fish Biol.*, **21**: 411-415.

ENK, M.D. & MATHIS, B.J. (1977) Distribution of cadmium and lead in a stream ecosystem. *Hydrobiologia*, **52**: 153-158.

FALCONER, C.R., DAVIES, I.M., & TOPPING, G. (1983) Trace metals in the common porpoise, *Phocoena phocoena*. *Mar. environ. Res.*, **8**: 119-127.

FINLEY, M.T. & DIETER, M.P. (1978) Influence of laying on lead accumulation in bone of mallard ducks. *J. Toxicol. environ. Health*, **4**: 123-128.

FINLEY, M.T., DIETER, M.P., & LOCKE, L.N. (1976) Lead in tissues of mallard ducks dosed with two types of lead shot. *Bull. environ. Contam. Toxicol.*, **16**: 261-269.

FISCHER, E., FILIP, J., MOLNAR, L., & NAGY, E. (1980) Karyometric studies of the effect of lead and cadmium in relation to the oxygen supply in the chloragocytes of *Tubifex tubifex* Muller. *Environ. Pollut.*, **21**: 203-207.

FRASER, J., PARKIN, D.T., & VERSPOOR, E. (1978) Tolerance to lead in the freshwater isopod *Asellus aquaticus*. *Water Res.*, **12**: 637-641.

FREDERICK, R.B. (1976) Effects of lead nitrate ingestion on open-field behaviour of mallard ducklings. *Bull. environ. Contam. Toxicol.*, **16**: 739-742.

FREEDMAN, M.L., CUNNINGHAM, P.M., SCHINDLER, J.E., & ZIMMERMAN, M.J. (1980) Effect of lead speciation on toxicity. *Bull. environ. Contam. Toxicol.*, **25**: 389-393.

GETZ, L.L., BEST, L.B., & PRATHER, M. (1977) Lead in urban and rural song birds. *Environ. Pollut.*, **12**: 235-238.

GIATTINA, J.D. & GARTON, R.R. (1983) A review of the preference-avoidance responses of fishes to aquatic contaminants. *Residue Rev.*, **87**: 43-90.

GILES, F.E., MIDDLETON, S.G., & GRAU, J.G. (1973) Evidence for the accumulation of atmospheric lead by insects in areas of high traffic density. *Environ. Entomol.*, **2**: 299-300.

GILL, J.M., HUGUET, J.H., & PEARSON, E.A. (1960) Submarine dispersal system for treated chemical wastes. *J. Water Pollut. Control Fed.*, **32**: 858-867.

GILMARTIN, M. & REVELANTE, N. (1975) The concentration of mercury, copper, nickel, silver, cadmium, and lead in the northern Adriatic anchovy, *Engraulis encrasicholus*, and sardine, *Sardina pilchardus*. *Fish. Bull.*, **73**: 193-201.

GOLDSMITH, C.D. & SCANLON, P.F. (1977) Lead levels in small mammals and selected invertebrates associated with highways of different traffic densities. *Bull. environ. Contam. Toxicol.*, **17**: 311-316.

GRANDY, J.W., LOCKE, L.N., & BAGLEY, G.E. (1968) Relative toxicity of lead and five proposed substitute shot types to pen-reared mallards. *J. wildl. Manage.*, **32**: 483-488.

GRAY, J.S. & VENTILLA, R.T. (1971) Pollution effects on micro and meifauna of sand. *Mar. Pollut. Bull.*, **2**: 39-43.

GRAY, J.S. & VENTILLA, R.T. (1973) Growth rates of sediment-living marine protozoans as a toxicity indicator for heavy metals. *Ambio*, **2**: 118-121.

GRUE, C.E., O'SHEA, T.J., & HOFFMAN, D.J. (1984) Lead concentrations and reproduction in highway-nesting barn swallows. *Condor*, **86**: 383-389.

GRUE, C.E., HOFFMAN, D.J., BEYER, W.N., & FRANSON, L.P. (1986) Lead concentrations and reproductive success in European starlings *Sturnus vulgaris* nesting within highway roadside verges. *Environ. Pollut.*, **42**: 157-182.

HAEGELE, M.A. & TUCKER, R.K. (1974) Effects of 15 common environmental pollutants on eggshell thickness in mallards and *Coturnix*. *Bull. environ. Contam. Toxicol.*, **11**: 98-102.

HAEGELE, M.A., TUCKER, R.K., & HUDSON, R.H. (1974) Effects of dietary mercury and lead on eggshell thickness in mallards. *Bull. environ. Contam. Toxicol.*, **11**: 5-11.

HALE, J.G. (1977) Toxicity of metal mining wastes. *Bull. environ. Contam. Toxicol.*, **17**: 66-73.

HARTENSTEIN, R., NEUHAUSER, E.F., & COLLIER, J. (1980) Accumulation of heavy metals in the earthworm *Eisenia foetida*. *J. environ. Qual.*, **9**: 23-26.

HEAD, P.C., D'ARCY, B.J., & OSBALDESTON, P.J. (1980) *The Mersey estuary bird mortality Autumn-Winter 1979 - preliminary report*, Warrington, United Kingdom, North West Water Authority, Directorate of Scientific Services (Scientific Report No. DSS-EST-80-1).

HEMPHILL, D.D. & RULE, J.H. (1975) Foliar uptake and translocation of ^{210}Pb and ^{109}Cd by plants. In: Hutchinson, T.O., ed. *Proceedings of the International Conference on Heavy Metals in the Environment, Toronto, October 1975*, Vol. III, pp. 77-86.

HESSLER, A. (1974) The effects of lead on algae. I. Effects of Pb on viability and mortality of *Platymonas subcaudiformis (Chlorophyta: Volvocales)*. *Water Air Soil Pollut.*, **3**: 371-385.

HESSLER, A. (1975) The effects of lead on algae. II. Mutagenesis experiments on *Platymonas subcordiformis* (Chlorophyta: Volvocales). *Mutat. Res.*, **31**: 43-47.

HILL, E.F. & CAMARDESE, M.B. (1986) *Lethal dietary toxicities of environmental contaminants and pesticides to Coturnix*, Washington, DC, US Department of the Interior, Fish and Wildlife Service, pp. 86-88 (Fish and Wildlife Technical Report No. 2).

HINDLE, E. (1932) A new kidney virus. *Nature (Lond.)*, **129**: 796.

HINDLE, E. & STEVENSON, A.C. (1930) Hitherto undescribed intranuclear bodies in the wild rat and monkeys, compared with known virus bodies in other animals. *Trans. R. Soc. Trop. Med. Hyg.*, **23**: 327.

HODSON, P.V. (1976) Delta-amino levulinic acid dehydratase activity of fish blood as an indicator of a harmful exposure to lead. *J. Fish. Res. Board Can.*, **33**: 268-271.

HODSON, P.V., BLUNT, B.R., & SPRY, D.J. (1978a) Chronic toxicity of water-borne and dietary lead to rainbow trout *(Salmo gairdneri)* in Lake Ontario water. *Water Res.*, **12**: 869-878.

HODSON, P.V., BLUNT, B.R., & SPRY, D.J. (1978b) pH-induced changes in blood of lead-exposed rainbow trout *(Salmo gairdneri)*. *J. Fish. Res. Board Can.*, **35**: 437-445.

HODSON, P.V., BLUNT, B.R., SPRY, D.J., & AUSTIN, K. (1977) Evaluation of erythrocyte delta-amino levulinic acid dehydratase activity as a short-term indicator in fish of a harmful exposure to lead. *J. Fish. Res. Board Can.*, **34**: 501-508.

HODSON, P.V., HILTON, J.W., BLUNT, B.R., & SLINGER, S.J. (1980) Effects of dietary ascorbic acid on chronic lead toxicity to young rainbow trout *(Salmo gairdneri)*. *Can. J. Fish. aquat. Sci.*, **37**: 170-176.

HOFFMAN, D.J., FRANSON, J.C., PATTEE, O.H., BUNCK, C.N., & ANDERSON, A. (1985a) Survival, growth and accumulation of ingested lead in nestling American kestrels *(Falco sparverius)*. *Arch. environ. Contam. Toxicol.*, **14**: 89-94.

HOFFMAN, D.J., FRANSON, J.C., PATTEE, O.H., BUNCK, C.M., & MURRAY, H.C. (1985b) Biochemical and hematological effects of lead ingestion in nestling American kestrels *(Falco sparverius)*. *Comp. Biochem. Physiol.*, **80**: 431-439.

HOLCOMBE, G.W., BENOIT, D.A., LEONARD, E.N., & MCKIM, J.M. (1976) Long-term effects of lead exposure on three generations of brook trout (*Salvelinus fontinalis). J. Fish. Res. Board Can.*, **33**: 1731-1741.

HOLL, W. & HAMPP, R. (1975) Lead and plants. *Residue Rev.*, **54**: 79-111.

HONDA, K., FUJISE, Y., TATSUKAWA, R., ITANA, K., & MIYAZAKI, N. (1986) Age-related accumulation of heavy metals in bone of the striped dolphin, *Stenella coeruleoalba. Mar. environ. Res.*, **20**: 143-160.

HONGVE, D., SKOGHEIM, O.K., HINDER, A., & ABRAHAMSEN, H. (1980) Effects of heavy metals in combination with NTA, humic acid, and suspended sediment on natural phytoplankton photosynthesis. *Bull. environ. Contam. Toxicol.*, **25**: 594-600.

HOOPER, M.C. (1937) An investigation of the effect of lead on plants. *Ann. appl. Biol.*, **24**: 690-695.

HOWELL, R. (1984) Acute toxicity of heavy metals to two species of marine nematodes. *Mar. environ. Res.*, **11**: 153-161.

HUDSON, R.H., TUCKER, R.K., & HAEGELE, M.A. (1984) *Handbook of toxicity of pesticides to wildlife*, 2nd ed., Washington, DC, US Department of the Interior, Fish and Wildlife Service, p. 80 (Resource Publication No. 153).

HUTTON, M. (1980) Metal contamination of feral pigeons *Columba livia* from the London area: Part II. Biological effects of lead exposure. *Environ. Pollut.*, **22**: 281-293.

HUTTON, M. & GOODMAN, G.T. (1980) Metal contamination of feral pigeons *Columbia livia* from the London area: Part I. Tissue accumulation of lead, cadmium, and zinc. *Environ. Pollut.*, **22**: 207-217.

IRELAND, M.P. (1977) Lead retention in toads *Xenopus laevis* fed increasing levels of lead-contaminated earthworms. *Environ. Pollut.*, **12**: 85-92.

IRWIN, J.C. & KARSTAD, L.H. (1972) The toxicity for ducks of disintegrated lead shot in a simulated-marsh environment. *J. wildl. Dis.*, **8**: 149-154.

JACKIM, E. (1973) Influence of lead and other metals on fish delta-aminolevulinate dehydrase activity. *J. Fish. Res. Board Can.*, **30**: 560-562.

JEFFERIES, D.J. & FRENCH, M.C. (1972) Lead concentrations in small mammals trapped on roadside verges and field sites. *Environ. Pollut.*, **3**: 147-156.

JOHANSSON-SJOBECK, M.L. & LARSSON, A. (1979) Effects of inorganic lead on delta-aminolevulinic acid dehydratase activity and hematological variables in the rainbow trout, *Salmo gairdneri*. *Arch. environ. Contam. Toxicol.*, **8**: 419-431.

JOHNSON, M.S., ROBERTS, R.D., HUTTON, M., & INSKIP, M.J. (1978) Distribution of lead, zinc and cadmium in small mammals from polluted environments. *Oikos*, **30**: 153-159.

JOHNSON, W.L. & DAMRON, B.L. (1982) Influence of lead acetate or lead shot ingestion upon white Chinese geese. *Bull. environ. Contam. Toxicol.*, **29**: 177-183.

JONES, J.R.E. (1938) The relative toxicity of salts of lead, zinc, and copper to the stickleback. *J. exp. Biol.*, **15**: 394-407.

JONES, J.R.E. (1948) A further study of the reactions of fish to toxic solutions. *J. exp. Biol.*, **25**: 22.

JOWETT, D. (1958) Populations of *Agrostis* spp. tolerant of heavy metals. *Nature (Lond.)*, **182**: 816-817.

KAPLAN, H.M., ARNHOLT, T.J., & PAYNE, J.E. (1967) Toxicity of lead nitrate solutions for frogs *(Rana pipiens)*. *Lab. Anim. Care*, **17**: 240-246.

KAY, S.H. & HALLER, W.T. (1986) Heavy metal bioaccumulation and effects on waterhyacinth weevils, *Neochetina eichhorniae*, feeding on waterhyacinth, *Eichhornia crassipes*. *Bull. environ. Contam. Toxicol.*, **37**: 239-245.

KAY, S.H., HALLER, W.T., & GARRARD, L.A. (1984) Effects of heavy metals on water hyacinths *(Eichhornia crassipes* (Mart.) Solms.) *Aquat. Toxicol.*, **5**: 117-128.

KEATON, C.M. (1937) The influence of lead compounds on the growth of barley. *Soil Sci.*, **43**: 401-411.

KENDALL, R.J. & SCANLON, P.F. (1982) The toxicology of ingested lead acetate in ringed turtle doves *Streptopelia risoria*. *Environ. Pollut.*, **27**: 255-262.

KHALID, B.Y., SALIH, B.M., & ISSAC, M.W. (1981) Lead contamination of soil in Baghdad city, Iraq. *Bull. environ. Contam. Toxicol.*, **27**: 634-638.

KHANGAROT, B.S. & RAY, P.K. (1987) Correlation between heavy metal acute toxicity values in *Daphnia magna* and fish. *Bull. environ. Contam. Toxicol.*, **38**: 722-726.

KHANGAROT, B.S., SEHGAL, A., & BHASIN, M.K. (1985) "Man and biosphere" - Studies on the Sikkim Himalayas. Part 5: Acute toxicity of selected heavy metals on the tadpoles of *Rana hexadactyla*. *Acta hydrochim. hydrobiol.*, **13**: 259-263.

KILHAM, L., LOW, R.J., CONTI, S.F., & DALLENBACK, F.D. (1962) Intranuclear inclusions and neoplasms in the kidneys of wild rats. *J. Natl Cancer Inst.*, **29**: 863-885.

KRISHNAJA, A.P., REGE, M.S., & JOSHI, A.G. (1987) Toxic effects of certain heavy metals (Hg, Cd, Pb, As and Se) on the intertidal crab *Scylla serrata*. *Mar. environ. Res.*, **21**: 109-119.

LAGERWERFF, J.V., ARMIGER, W.H., & SPECHT, A.W. (1973) Uptake of lead by alfalfa and corn from soil and air. *Soil Sci.*, **115**: 455-460.

LANE, S.D. & MARTIN, E.S. (1977) A histochemical investigation of lead uptake in *Raphanus sativus*. *New Phytol.*, **79**: 281-286.

LEWIS, J.C. & LEGLER, E. (1968) Lead shot ingestion by mourning doves and incidence in soil. *J. wildl. Manage.*, **32**: 476-482.

LEWIS, T.E. & MCINTOSH, A.W. (1986) Uptake of sediment-bound lead and zinc by the freshwater isopod *Asellus communis* at three different pH levels. *Arch. environ. Contam. Toxicol.*, **15**: 495-504.

LLOYD, R. (1961) Effect of dissolved oxygen concentration on the toxicity of several poisons to rainbow trout *(Salmo gairdneri* Richardson). *J. exp. Biol.*, **38**: 447-455.

LOCKE, L.N. & BAGLEY, G.E. (1967) Lead poisoning in a sample of Maryland mourning doves. *J. wildl. Manage.*, **31**: 515-518.

LU, P.Y., METCALF, R.L., FURMAR, R., VOGEL, R., & HASSETT, J. (1975) Model ecosystem studies of lead and cadmium and of urban sewage sludge containing these elements. *J. environ. Qual.*, **4**: 505-509.

MADDOCK, B.G. & TAYOR, D. (1980) The acute toxicity and bioaccumulation of some lead alkyl compounds in marine animals. In: Branica, M. & Konrad, Z., ed. *Lead in the marine environment*, Oxford, Pergamon Press, pp. 233-261.

MALANCHUK, J.L. & GRUENDLING, G.K. (1973) Toxicity of lead nitrate to algae. *Water Air Soil Pollut.*, **2**: 181-190.

MALONE, C., KOEPPE, D.E., & MILLER, R.J. (1974) Localization of lead accumulated by corn plants. *Plant Physiol.*, **53**: 388-394.

MARCHETTI, R. (1978) Acute toxicity of alkyl leads to some marine organisms. *Mar. Pollut. Bull.*, **9**: 206-207.

MARTIN, W.E. (1972) Mercury and lead residues in starlings - 1970. *Pestic. monit. J.*, **6**: 27-32.

MARTIN, W.E. & NICKERSON, P.R. (1973) Mercury, lead, cadmium, and arsenic residues in starlings - 1971. *Pestic. monit. J.*, **7**: 67-72.

MAUTINO, M. & BELL, J.U. (1987) Hematological evaluation of lead intoxication in mallards. *Bull. environ. Contam. Toxicol.*, **38**: 78-85.

MAY, T.W. & MCKINNEY, G.L. (1981) Cadmium, lead, mercury, arsenic, and selenium concentrations in freshwater fish, 1976-77. National Pesticide Monitoring Program. *Pestic. monit. J.*, **15**: 14-38.

MAYER, F.L. & ELLERSIECK, M.R. (1986) *Manual of acute toxicity: interpretation and data base for 410 chemicals and 66 species of freshwater animals*, Washington, DC, US Department of the Interior, Fish and Wildlife Service, 506 pp (Resource Publication No. 160).

MAYES, R.A., MCINTOSH, A.W., & ANDERSON, V.L. (1977) Uptake of cadmium and lead by a rooted aquatic macrophyte *(Elodea canadensis)*. *Ecology*, **58**: 1176-1180.

MERLINI, M. & POZZI, G. (1977) Lead and freshwater fishes Part I. Lead accumulation and water pH. *Environ. Pollut.*, **12**: 167-172.

MEYER, W., HARISCH, G., & SAGREDOS, A.N. (1986) Biochemical and histochemical aspects of lead exposure in dragonfly larvae *(Odonata: Anisoptera)*. *Ecotoxicol. environ. Saf.*, **11**: 308-319.

MIERAU, G.W. & FAVARA, B.E. (1975) Lead poisoning in roadside populations of deer mice. *Environ. Pollut.*, **8**: 55-64.

MONAHAN, T.J. (1976) Lead inhibition of chlorophycean microalgae. *J. Phycol.*, **12**: 358-362.

MORGAN, G.W., EDENS, F.W., THAXTON, P., & PARKHURST, C.R. (1975) Toxicity of dietary lead in Japanese quail. *Poult. Sci.*, **54**: 1636-1642.

MUDGE, G.P. (1983) The incidence and significance of lead pellet poisoning in British wildfowl. *Biol. Conserv.*, **27**: 333-372.

MUDRE, J.M. & NEY, J.J. (1986) Patterns of accumulation of heavy metals in the sediment of roadside streams. *Arch. environ. Contam. Toxicol.*, **15**: 489-493.

MURAMOTO, S. (1980) Effect of complexans (EDTA, NTA and DTPA) on the exposure to high concentrations of cadmium, copper, zinc and lead. *Bull. environ. Contam. Toxicol.*, **25**: 941-946.

MURAMOTO, S. & OKI, Y. (1983) Removal of some heavy metals from polluted water by water hyacinth *(Eichhornia crassipes). Bull. environ. Contam. Toxicol.*, **30**: 170-177.

NAKADA, M., FUKAYA, K., TOKESHITA, S., & WADA, Y. (1979) The accumulation of heavy metals in the submerged plant *(Elodea nuttallii). Bull. environ. Contam. Toxicol.*, **22**: 21-27.

NCC (1981) *Lead poisoning in swans*, London, Nature Conservancy Council.

NEHRING, B. (1976) Aquatic insects as biological monitors of heavy metal pollution. *Bull. environ. Contam. Toxicol.*, **15**: 147-154.

NEY, J.J. & VAN HASSEL, J.H. (1983) Sources of variability in accumulation of heavy metals by fishes in a roadside stream. *Arch. environ. Contam. Toxicol.*, **12**: 701-706.

OBERLANDER, H.E. & ROTH, K. (1978) [Effect of the heavy metals chromium, nickel, copper, zinc, cadmium, mercury and lead on uptake and translocation of potassium and phosphate by young barley plants.] *Z. Pflanzenernaehr. Bodenkd.*, **141**: 107-116 (in German).

OHI, G., HIRONOBU, S., AKIYAMA, K., & YAGYU, H. (1974) The pigeon, a sensor of lead pollution. *Bull. environ. Contam. Toxicol.*, **12**: 92-98.

OSBORN, D. (1979) Seasonal changes in the fat, protein and metal content of the liver of the starling *(Sturnus vulgaris). Environ. Pollut.*, **19**: 145-155.

OSBORN, D., EVERY, W.J., & BULL, K.R. (1983) The toxicity of trialkyl lead compounds to birds. *Environ. Pollut.*, **31**: 261-275.

OZOH, P.T.E. (1979) Malformations and inhibitory tendencies induced to *Brachydanio rerio* (Hamilton-Buchanan) eggs and larvae due to exposures in low concentrations of lead and copper ions. *Bull. environ. Contam. Toxicol.*, **21**: 668-675.

PATTEE, O.H. (1984) Eggshell thickness and reproduction in American kestrels exposed to chronic dietary lead. *Arch. environ. Contam. Toxicol.*, **13**: 29-34.

PATTEE, O.H., WEIMEYER, S.N., MULHERN, B.M., SILEO, L., & CARPENTER, M. (1981) Experimental lead shot poisoning in bald eagles. *J. wildl. Manage.*, **45**: 806-810.

PERSOONE, G. & UYTTERSPROT, G. (1975) The influence of inorganic and organic pollutants on the rate of reproduction of a marine hypotrichous ciliate: *Euplotes vannus* Muller. *Rev. int. Océanogr. méd.*, **37-38**: 125-151.

PERTTILLA, M., TERVO, V., & PARMANNE, R. (1982) Age-dependence of the concentrations of harmful substances in Baltic herring *(Clupea harengus)*. *Chemosphere*, **11**: 1019-1026.

PETERSON, P.J. (1978) Lead and vegetation. In: Nriagu, J.O., ed. *Biochemistry of lead in the environment*, Amsterdam, Oxford, New York, Elsevier Science Publishers, pp. 357-384.

PICKERING, Q.H. & HENDERSON, C. (1966) The acute toxicity of some heavy metals to different species of warmwater fishes. *Air Water Pollut. int. J.*, **10**: 453-463.

PORTMANN, J.E. & WILSON, K.W. (1971) The toxicity of 140 substances to the brown shrimp and other marine animals. *MAFF Shellfish Inf. Leafl.*, 22: 1-11.

PRASAD, P.V.D. & PRASAD, P.S.D. (1982) Effect of cadmium, lead, and nickel on three freshwater green algae. *Water Air Soil Pollut.*, **17**: 263-268.

PRICE, P.W., RATHCKE, B.J., & GENTRY, D.A. (1974) Lead in terrestrial arthropods: Evidence for biological concentration. *Environ. Entomol.*, **3**: 370-372.

PRINGLE, B.H., HISSONG, D.E., KATZ, E.L., & MULAWKA, S.T. (1968) Trace metal accumulation by estuarine mollusks. *J. sanit. Eng. Div., Proc. Am. Soc. Civil Eng.*, **94**(SA3): 455-475.

QUARLES, H.O., HANAWALT, R.B., & ODUM, W.E. (1974) Lead in small mammals, plants, and soil at varying distances from a highway. *J. appl. Ecol.*, **11**: 937-950.

RAINS, D.W. (1971) Lead accumulation by wild oats *(Avena fatua)* in a contaminated area. *Nature (Lond.)*, **233**: 210-211.

RAY, S., MCLEESE, D.W., & PERTERSON, M.R. (1981) Accumulation of copper, zinc, cadmium and lead from two contaminated sediments by three marine invertebrates - A laboratory study. *Bull. environ. Contam. Toxicol.*, **26**: 315-322.

REISH, D.J., MARTIN, J.M., PILTZ, F.M., & WORD, J.Q. (1976) The effect of heavy metals on laboratory populations of two polychaetes with comparisons to water quality conditions and standards in southern California marine waters. *Water Res.*, **10**: 299-302.

RIDGWAY, L.P. & KARNOFSKY, D.A. (1952) The effects of metals on the chick embryo: toxicity and production of abnormalities in development. *Ann. N.Y. Acad. Sci.*, **55**: 203-215.

ROBERTS, D. & MAGUIRE, C. (1976) Interactions of lead with sediments and meiofauna. *Mar. Pollut. Bull.*, **7**: 211-214 .

ROBERTS, R.D., JOHNSON, M.S., & HUTTON, M. (1978) Lead contamination of small mammals from abandoned metalliferous mines. *Environ. Pollut.*, **15**: 61-69.

RODERER, G. (1980) On the toxic effects of tetraethyl lead and its derivatives on the chrysophyte *Poteriochromonas malhamensis*. I. Tetraethyl lead. *Environ. Res.*, **23**: 371-384.

RODERER, G. (1983) On the toxic effects of tetraethyl lead and its derivatives on the chrysophyte *Poteriochromonas malhamensis*. IV. Influence of lead antidotes and related agents. *Chem.-biol. Interact.*, **48**: 247-254.

RODERER, G. (1986) On the toxic effects of tetraethyl lead and its derivatives on the chrysophyte *Poteriochromonas malhamensis*. VII. Protective action of thiol compounds, vitamins, trace elements, and other agents. *Ecotoxicol. environ. Saf.*, **11**: 277-294.

ROSENWEIG, W. & PRAMER, D. (1980) Influence of cadmium, zinc, and lead on growth, trap formation and collagenase activity of nematode-trapping fungi. *Appl. environ. Microbiol.*, **40**: 694-696.

ROZMAN, R.S., LOCKE, L.N., & MCCLURE, S.F. (1974) Enzyme changes in mallard ducks fed iron or lead shot. *Avian Dis.*, **18**: 435-445.

RUHLING, A. & TYLER, G. (1970) Regional differences in the deposition of heavy metals over Scandinavia. *J. appl. Ecol.*, **8**: 497-507.

RUTHVEN, J.A. & CAIRNS, J. (1973) Response of fresh-water protozoan artificial communities to metals. *J. Protozool.*, **20**: 127-135.

SASTRY, K.V. & GUPTA, P.K. (1978a) Alterations in the activity of some digestive enzymes of *Channa punctatus* exposed to lead nitrate. *Bull. environ. Contam. Toxicol.*, **19**: 549-555.

SASTRY, K.V. & GUPTA, P.K. (1978b) *In vitro* inhibition of digestive enzymes by heavy metals and their reversal by a chelating agent. Lead nitrate intoxication. *Bull. environ. Contam. Toxicol.*, **20**: 736-742.

SCHAFER, E.W., BOWLES, W.A., & HURLBUT, J. (1983) The acute oral toxicity, repellency, and hazard potential of 998 chemicals to one or more species of wild and domestic birds. *Arch. environ. Contam. Toxicol.*, **12**: 355-382.

SHAFFI, S.A. (1979) Lead toxicity, biochemical and physiological imbalance in nine freshwater teleosts. *Toxicol. Lett.*, **4**: 155-161.

SILVERBERG, B.A., WONG, P.T.S., & CHAU, Y.K. (1977) Effect of tetramethyl lead on freshwater green algae. *Arch. environ. Contam. Toxicol.*, **5**: 305-313.

SIMPSON, V.R., HUNT, A.E., & FRENCH, M.C. (1979) Chronic lead poisoning in a herd of mute swans. *Environ. Pollut.*, **18**: 187-202.

SMITH, G.J. & RONGSTAD, O.J. (1982) Small mammal heavy metal concentrations from mined and control sites. *Environ. Pollut.*, **28**: 121-134.

SOMERO, G.N., CHOW, T.J., YANCEY, P.H., & SNYDER, C.B. (1977) Lead accumulation rates in tissues of the estuarine teleost fish, *Gillichthys mirabilis*: salinity and temperature effects. *Arch. environ. Contam. Toxicol.*, **6**: 337-348.

SPEHAR, R.L., ANDERSON, R.L., & FIANDT, J.T. (1978) Toxicity and bioaccumulation of cadmium and lead in aquatic invertebrates. *Environ. Pollut.*, **15**: 195-208.

STANLEY, R.A. (1974) Toxicity of heavy metals and salts to Eurasian watermilfoil *(Myriophyllum spicatum* L.). *Arch. environ. Contam. Toxicol.*, **2**: 331-341.

STENDELL, R.C. (1980) Dietary exposure of kestrels to lead. *J. wildl. Manage.*, **44**: 527-530.

STROMGREN, T. (1982) Effect of heavy metals (zinc, mercury, copper, cadmium, lead, nickel) on the length growth of *Mytilus edulis*. *Mar. Biol.*, **72**: 69-72.

SUBBAIAH, M.B., NAIDU, K.A., PURUSHOTHAM, K.R., & RAMAMURTHI, R. (1983) Heavy metal toxicity to some freshwater organisms. *Geobios*, **10**: 128-129.

SUMMERFELT, R.C. & LEWIS, W.M. (1967) Repulsion of green sunfish by certain chemicals. *J. Water Pollut. Control Fed.*, **39**: 2030.

SYVESTON, J.T. & LARSON, C.L. (1947) Intranuclear inclusion bodies in the kidneys of wild rats. *Arch. Pathol.*, **43**: 541-552.

TAYLOR, D., MADDOCK, B.G., & MANCE, G. (1985) The acute toxicity of nine "grey list" metals (arsenic, boron, chromium, copper, lead, nickel, tin, vanadium and zinc) to two marine fish species: dab (*Limanda limanda*) and grey mullet (*Chelon labrosus*). *Aquat. Toxicol.*, **7**: 135-144.

TIRAVANTI, G. & BOARI, G. (1979) Potential pollution of a marine environment by lead alkyls: the cavtat incident. *Environ. Sci. Technol.*, **13**: 849-854.

TORNABENE, T.G. & EDWARDS, H.W. (1972) Microbial uptake of lead. *Science*, **176**: 1334-1335.

TORNABENE, T.G. & EDWARDS, H.W. (1973) Effects of lead on bacterial membranes. In: *Proceedings of the 7th Annual Conference on Trace Substances in Environmental Health*, Missouri, Columbia, University of Missouri Press.

TORNABENE, T.G. & PETERSON, S.L. (1975) Interaction of lead and bacterial lipids. *Appl. Microbiol.*, **29**: 680-684.

TRAINER, D.O. & HUNT, R.A. (1965) Lead poisoning of waterfowl in Wisconsin. *J. wildl. Manage.*, **29**: 95-103.

TURNBULL, H., DE MANN, J.G., & WESTON, R.F. (1954) Toxicity of various refinery materials to frehwater fish, Symposium on Waste Disposal in the petroleum Industry. *Ind. Eng. Chem.*, **46**: 324-333.

VAN DER WERFF, M. & PRUYT, M.J. (1982) Long-term effects of heavy metals on aquatic plants. *Chemosphere*, **11**: 727-739.

VAN HASSEL, J.H., NEY, J.J., & GARLING, D.L. (1979) Seasonal variations in the heavy metal concentrations of sediments influenced by highway of different traffic volumes. *Bull. environ. Contam. Toxicol.*, **23**: 592-596.

VAN HASSEL, J.H., NEY, J.J., & GARLING, D.L. (1980) Heavy metals in a stream ecosystem at sites near highways. *Trans. Am. Fish. Soc.*, **109**: 636-643.

VAN HOOK, R.I. (1974) Cadmium, lead, and zinc distributions between earthworms and soils: potentials for biological accumulation. *Bull. environ. Contam. Toxicol.*, **12**: 509-512.

VAN STRAALEN, N.M. & VAN MEERENDONK, J.H. (1987) Biological half-lives of lead in *Orchesella cincta* (L.) (Collembola). *Bull. environ. Contam. Toxicol.*, **38**: 213-219.

VARANASI, U., ROBISCH, P.A., & MALINS, D.C. (1975) Structural alterations in fish epidermal mucus produced by water-borne lead and mercury. *Nature (Lond.)*, **258**: 431-432.

VENGRIS, V.E. & MARE, C.J. (1974) Lead poisoning in chickens and the effect of lead on interferon and antibody production. *Can. J. comp. Med.*, **38**: 328-335.

VIGHI, M. (1981) Lead uptake and release in an experimental chain. *Ecotoxicol. environ. Saf.*, **5**: 177-193.

WAINWRIGHT, S.J. & WOOLHOUSE, H.W. (1975) Physiological mechanisms of heavy metal tolerance. In: Chadwick, M.J. & Goodman, G.T., ed. *The ecology of resource degradation and renewal*, Oxford, Blackwell Scientific Publications.

WALLEN, I.E., GREER, W.C., & LASATER, R. (1957) Toxicity to *Gambusia affinis* of certain pure chemicals in turbid waters. *Sewage ind. Wastes*, **29**: 695-711.

WARD, N.I., REEVES, R.D., & BROOKS, R.R. (1975) Lead in soil and vegetation along a New Zealand state highway with low traffic volume. *Environ. Pollut.*, **9**: 243-251.

WARNICK, S.L. & BELL, H.L. (1969) The acute toxicity of some heavy metals to different species of aquatic insects. *J. Water Pollut. Control Fed.*, **41**: 280-284.

WATLING, H.R. (1983a) Accumulation of seven metals by *Crassostrea gigas, Crassostrea margaritacea, Perna perna, and Choromytilus meridionalis. Bull. environ. Contam. Toxicol.*, **30**: 317-322.

WATLING, H.R. (1983b) Comparative study of the effects of metals on the settlement of *Crassostrea gigas. Bull. environ. Contam. Toxicol.*, **31**: 344-351.

WEIR, A. & HINE, C.H. (1970) Effects of various metals on behavior of conditioned goldfish. *Arch. environ. Health*, **20**: 45-51.

WEIS, J.S. & WEIS, P. (1977) Effects of heavy metals on development of the killifish, *Fundulus heteroclitus. J. fish Biol.*, **11**: 49-54.

WEISMAN, L. & SKROBAK, J. (1980) Toxicity of food with increased content of lead for the caterpillar *Scotia segetum. Biologia (Bratislava)*, **35**: 823-826.

WEISMAN, L. & SVATARAKOVA, L. (1981) The influence of lead on some vital manifestations of insects. *Biologia (Bratislava)*, **36**: 147-151.

WELCH, W.R. & DICK, D.L. (1975) Lead concentrations in tissues of roadside mice. *Environ. Pollut.*, **8**: 15-21.

WHEELER, G.L. & ROLFE, G.L. (1979) The relationship between daily traffic volume and the distribution of lead in roadside soil and vegetation. *Environ. Pollut.*, **18**: 265-274.

WHITTON, B.A. (1970) Toxicity of zinc, copper and lead to chlorophyta from flowing waters. *Arch. Mikrobiol.*, **72**: 353-360.

WHO (1977) *Environmental Health Criteria 3: Lead*, Geneva, World Health Organization, 160 pp.

WILBER, C.G. (1969) *The biological effects of water pollution*, Springfield, Illinois, C.C. Thomas.

WILLIAMSON, P. & EVANS, P.R. (1972) Lead: Levels in roadside invertebrates and small mammals. *Bull. environ. Contam. Toxicol.*, **8**: 280-288.

WONG, P.T.S., CHAU, Y.K., KRAMAR, O., & BENGERT, G.A. (1981) Accumulation and depuration of tetramethyllead by rainbow trout. *Water Res.*, **15**: 621-625.

ZIMDAHL, R.L., MCCREARY, D.T., & GWYNN, S.M. (1978) Lead uptake by plants - the influence of lead source. *Bull. environ. Contam. Toxicol.*, **19**: 431-435.

ZOOK, B.C., SAUER, R.M., & GARNER, F.M. (1972) Lead poisoning in captive wild animals. *J. wildl. Dis.*, **8**: 264-272.

Other titles available in the ENVIRONMENTAL HEALTH CRITERIA series (continued):

44. Mirex
45. Camphechlor
46. Guidelines for the Study of Genetic Effects in Human Populations
47. Summary Report on the Evaluation of Short-term Tests for Carcinogens (Collaborative Study on *In Vitro* Tests)
48. Dimethyl Sulfate
49. Acrylamide
50. Trichloroethylene
51. Guide to Short-term Tests for Detecting Mutagenic and Carcinogenic Chemicals
52. Toluene
53. Asbestos and Other Natural Mineral Fibres
54. Ammonia
55. Ethylene Oxide
56. Propylene Oxide
57. Principles of Toxicokinetic Studies
58. Selenium
59. Principles for Evaluating Health Risks from Chemicals During Infancy and Early Childhood: The Need for a Special Approach
60. Principles and Methods for the Assessment of Neurotoxicity Associated With Exposure to Chemicals
61. Chromium
62. 1,2-Dichloroethane
63. Organophosphorus Insecticides - A General Introduction
64. Carbamate Pesticides - A General Introduction
65. Butanols - Four Isomers
66. Kelevan
67. Tetradifon
68. Hydrazine
69. Magnetic Fields
70. Principles for the Safety Assessment of Food Additives and Contaminants in Food
71. Pentachlorophenol
72. Principles of Studies on Diseases of Suspected Chemical Etiology and Their Prevention
73. Phosphine and Selected Metal Phosphides
74. Diaminotoluenes
75. Toluene Diisocyanates
76. Thiocarbamate Pesticides - A General Introduction
77. Man-made Mineral Fibres
78. Dithiocarbamate Pesticides
79. Dichlorvos
80. Pyrrolizidine Alkaloids
81. Vanadium
82. Cypermethrin
83. DDT and its Derivatives - Environmental Aspects
84. 2,4-Dichlorophenoxyacetic Acid - Environmental Aspects